Morality for Humans

Morality for Humans

*Ethical Understanding from the
Perspective of Cognitive Science*

MARK JOHNSON

The University of Chicago Press   Chicago and London

The University of Chicago Press, Chicago 60637
The University of Chicago Press, Ltd., London
© 2014 by The University of Chicago
All rights reserved. Published 2014.
Paperback edition 2015
Printed in the United States of America

24 23 22 21 20 19 18 17 16 15   2 3 4 5 6

ISBN-13: 978-0-226-11340-1 (cloth)
ISBN-13: 978-0-226-32494-4 (paper)
ISBN-13: 978-0-226-11354-8 (e-book)
DOI: 10.7208/chicago/9780226113548.001.0001

Library of Congress Cataloging-in-Publication Data

Johnson, Mark, 1949– author.
　Morality for humans: ethical understanding from the perspective of cognitive science / Mark Johnson.
　　　pages cm
　　　Includes bibliographical references and index.
　　　ISBN 978-0-226-11340-1 (cloth: alk. paper)—
ISBN 978-0-226-11354-8 (e-book) 1. Ethics. 2. Cognitive science—Moral and ethical aspects.　I. Title.
　BJ45.5.J64 2014
　171'.7—dc23   2013025456

♾ This paper meets the requirements of ANSI/NISO Z39.48-1992 (Permanence of Paper).

*For Sandra McMorris Johnson,
the most kind and loving person I have known*

Contents

Preface ix
Introduction: The Need for Ethical Naturalism 1

1 Moral Problem-Solving as an Empirical Inquiry 28
2 Where Are Our Values Bred?—Sources of Moral Norms 48
3 Intuitive Processes of Moral Cognition 73
4 Moral Deliberation as Cognition, Imagination, and Feeling 89
5 The Nature of "Reasonable" Moral Deliberation 112
6 There Is No Moral Faculty 137
7 Moral Fundamentalism Is Immoral 163
8 The Making of a Moral Self 192

Acknowledgments 223

Notes 225   *References 239*   *Index 249*

Preface

In *Man's Search for Meaning* (1946), Viktor Frankl chronicles three years of the most despicable degradation and nearly unbearable suffering that he personally experienced and witnessed in concentration camps such as Auschwitz and Dachau. Frankl is justly famous for his attempts to explain to those of us who probably cannot ever comprehend the magnitude of the evil just what it was like to suffer through those horrors. But even more importantly, Frankl attempts to help us understand how people could endure such torture by holding on to their sense of meaning in life, even when there seemed to be absolutely no reason to sustain hope.

In a section where he discusses the psychology of the camp guards, Frankl asks how human beings could possibly inflict such terrible suffering on other humans, but he also acknowledges that, in the midst of all this daily cruelty, there were camp officials who were relatively kind and caring. After the liberation of a certain camp, he reports that it was discovered that the camp commander had actually paid out of his own pocket for medicines for his prisoners, and that he apparently "never once lifted his hand against any of us." The senior warden in that same camp, on the other hand, took sadistic pleasure in beating the prisoners whenever he could. Frankl concludes:

From all this we may learn that there are two races of men in this world, but only these two—the "race" of the decent man and the "race" of the indecent man. Both are found everywhere; they penetrate into all groups of society. . . . Life in a concentration camp tore open the human soul and exposed its depths. Is it surprising that in those depths we again found only human qualities which in their very nature were a mixture of good and evil? The rift dividing good from evil, which goes through all human beings, reaches into the lowest depths and becomes apparent even on the bottom of the abyss which is laid open by the concentration camp. (1946/2006, 137–38)

Frankl's simple taxonomy thus recognizes but two types of humans—the decent and the indecent. How can we distinguish the one from the other? Oftentimes, it is pretty clear who falls into which type, though there will always be borderline cases. If you read Frankl's account of life in the camps, you will see that to be decent means caring for other people, respecting them, helping them when they are most in need, and generally treating them as people who, like you, suffer, require physical and psychological nurturance, need love, and seek meaning in their lives.

No matter how we circumscribe the realm of decent behavior, there is no avoiding the fact that not all cultures will draw the lines in precisely the same way; therefore, the most we can hope for are some very general ideals of human comportment. Frankl explains, for example, that in the camps being decent would not require that you sacrifice your life for another person, but it certainly would require that you not intentionally harm them to get something for yourself, such as a piece of bread. There is understandably a great deal of gray area on the borderline between decency and indecency, and we cannot eradicate this ambiguity, either via the commandments of a holy God or the laws of universal moral reason. There can be no deductive inference from the concept of decency to a uniquely specified set of prescribed and proscribed moral behaviors.

It is not my primary concern in this book to defend Frankl's taxonomy of the decent versus the indecent. Rather, what I want to focus on is whether, in order to mark off moral right from wrong, we need to posit some transcendent source of absolute moral values or principles. Many people are utterly convinced that if it should turn out that there are no sources of absolute moral values and principles, then there would be no principled, intelligent way to distinguish the decent from the indecent, the right from the wrong, the good from the bad. I think this view is profoundly mistaken. I am going to argue that human beings do not have, and never had, access to any such absolute principles. I will support my argument with evidence from research on human cognition, appraisal, and deliberation. I will then argue that the absence of any such absolutes is no obstacle to our ability to intelligently sort the moral from the immoral. Decency, or any alleged moral value or standard, can be tied entirely to human needs, values, and cultural arrangements, without any reliance on notions such as the eternal, the transcendent, or the supernatural.

There is nothing in our processes of moral deliberation that requires anything mysterious, esoteric, or transcendent to justify our moral appraisals. Our notions of moral decency, which concern the kinds of persons we ought to strive to become and how we ought to treat others, are entirely *human* notions, rooted in human nature, human needs, human thought, human social interaction, and human desires for a meaningful and fulfilled life. We can articulate a psychologically and philosophically adequate account of moral cognition and values that makes it possible to justify our moral appraisals, such as giving reasons why it would be wrong to treat innocent people in certain ways and why it would be good of you to show care and consideration for their basic bodily and psychological well-being. I will argue that all we ever had when it comes to questions of moral justification is our modest ability to give our best reasons and to show what life could be like if it were to realize certain ideals of character and behavior. This was never a matter of ultimate justification, claims to moral certainty, or reliance on supposedly trans-human foundations of moral knowledge. We are stuck with being finite, fallible human creatures who have to navigate our morally problematic landscape under the guidance of our very limited imaginative intelligence.

An important dimension of my argument in this book will consist in showing that what I call *moral fundamentalism*—the positing of absolute moral values, principles, or facts—is cognitively indefensible, because it is dramatically out of touch with contemporary mind science. Even worse, moral fundamentalism is immoral, I shall argue, because it cuts off the very processes of intelligent moral inquiry that we most need if we hope to face our pressing ethical concerns. Moral fundamentalism is the very worst possible strategy for anyone who hopes to deal intelligently with their moral problems.

However, in addition to criticizing claims to moral certainty and absolute principles, I need to give a positive and constructive account of what the process of moral deliberation looks like from the perspective of the cognitive sciences. I will argue that good moral deliberation is a form of problem-solving, in which we imaginatively project possible courses of action available to us, in order to determine which imagined course best resolves our actual moral problem. This kind of situated, imaginative moral inquiry does not need absolute foundations, which is a good thing, since humans never really had, and cannot ever have, access to the absolute. Moral deliberation at its best is a process of reconstructing our experience in a way that resolves the morally problematic situation that is currently confronting us. Such a process involves the only reasonable notion of transcendence available to humans—namely, the ability to move beyond our current habits of thought and action to creatively remake some aspect of ourselves and our world. There is nothing about such a process that takes us out of our skins, as if we were somehow little gods capable of generating moral absolutes. Instead, imaginative moral deliberation is embedded, embodied, and enacted within our changing, malleable experience. All of this transformative activity is entirely human in every respect, without any trace of supernatural grounding or reliance on alleged capacities of pure reason or will. The morality that results is thus a "morality (fit) for humans."

Introduction: The Need for Ethical Naturalism

Many people believe that the only way to avoid a vicious, dog-eat-dog moral relativism is to affirm eternal and universal moral values and principles, values whose source must lie in something that transcends the finiteness and vicissitudes of human existence. I was nurtured and educated in just such a transcendent, absolutist view, but over the years, and with much emotional and intellectual turmoil, I lost my conviction in the moral fundamentalism that underlies this perspective. The more I studied the nature of human concepts, understanding, and reasoning, the more I came to recognize profound problems with the picture of experience, thought, and value presupposed by views of morality as transcendentally grounded. My engagement with cognitive science research on human meaning, conceptualization, and reasoning led me to the realization that our values, including our ethical standards and ideals, emerge from our embodied, interpersonal, culturally situated habitation of *our* world, and not from some transcendent realm.

Surprisingly, this realization did not lead me to moral relativism, but rather to a conception of moral standards as relatively stable, but always provisional and corrigible, norms. Moreover, it brought me to an understanding that the key to intelligent moral inquiry is an imaginative process of moral deliberation by which our experience is reconstructed to achieve growth of meaning and enriched possibilities for human flourishing. I came to regard this as a psychologically realistic *morality for humans*, by which I mean a morality appropriate for actual human beings, with their limited and fallible cognitive and emotional capacities.

Toward an Ethics Naturalized

In this book, I attempt to articulate a naturalistic approach to values and moral deliberation that seems to me compatible with the account of embodied, situated meaning and understanding that has emerged in the cognitive sciences over the past three decades. I will situate my position in relation to some of the more influential contemporary accounts of moral psychology, such as the views of Robert Hinde (2002), Antonio Damasio (2003), Marc Hauser (2006), Owen Flanagan (2007), Patricia Churchland (2011), Philip Kitcher (2011), and Jonathan Haidt (2012). Such an approach requires a radical rethinking of some of our most deeply entrenched views about moral judgment and deliberation. In particular, it requires us to abandon the idea of some allegedly "pure" practical emotion-free reason, along with the correlative idea of unconditional moral principles. It thus rejects any form of moral absolutism or moral fundamentalism as being incompatible with how human beings actually understand and reason.

My alternative to these misguided absolutisms is a conception of moral deliberation as a form of imaginative problem-solving. In addition to recent experimental research that identifies two different processes of moral cognition—one consisting of nonconscious, fast, affect-laden, intuitive appraisals, and the other a conscious, slow, reflective, principled after-the-fact justificatory form of reasoning—I argue that there is also an important place for reflective, critical, and imaginative moral deliberation. This third process of moral cognition is emotionally driven but yet subject to assessments of reasonableness. I will describe and explain this third process in light of recent developments in the cognitive sciences. My goal is to articulate an understanding of moral cognition that is fit for human beings as we know them, not as we might wish them to be when we are under the mesmerizing spell of a quest for an illusory moral certainty. A moral philosophy fit for humans will regard persons as embodied, culturally embedded, highly complex organisms that are capable of an imaginative process of moral problem-solving. The view that emerges from this naturalistic perspective is anti-absolutist and fallibilist, yet it can provide us with guidance about what kinds of persons we should strive to become and what kinds of world we should seek to realize.

THE NEED FOR ETHICAL NATURALISM

My approach to moral cognition is naturalistic. Abraham Edel gives a good summary of what is involved in a naturalistic approach to ethics of the sort I will be developing:

> Ethical Naturalism, or naturalistic ethics, regards morality as a phenomenon in the natural world to be understood through the many ways we study nature. Its general attitude is this-worldly, not otherworldly or non-worldly: morality functions to further human survival, maintain community, and regulate relations to keep them effective; it can improve as well as support institutions, give scope to human capacities, and shape ideals as directions of activity in goal-seeking. Where naturalistic ethics has an explicit metaphysics, it shares with materialism a regard for matter and its ways as a resource, a limitation, a determinant, but it traditionally rejects reductionism or dualistic assumptions that qualities of consciousness are outside the natural world. (2001, 1217)

I should say a word, at the outset, about how I understand the term "naturalistic." Common parlance sometimes mistakenly draws a sharp ontological distinction between *natural* events and processes, on the one hand, and *cultural* institutions and practices, on the other. The former are thought to be governed by causal necessity and therefore are studied by the methods of the natural sciences, whereas the latter are regarded as matters of human freedom and meaning, and therefore require special non-causal, interpretive methods of investigation.

On the view I will be developing, there is no basis for drawing a radical dualistic distinction between nature and culture, as though each person had a "natural" (bodily, physical) self and a distinct and different "cultural" (social, moral) self that somehow have to coexist and interrelate. Culture is not a superficial veneer of shared meanings, values, and practices that are merely layered on top of some supposedly purely material organic being. Our nature as biological organisms is intricately intertwined with our cultural being, which, in turn, cannot be realized without biological creatures to enact it. It is thus part of our human nature that we live, move, and realize our being as at once both biological and cultural creatures. There are occasions when we find it useful to focus primarily on our biological characteristics, and there are plenty of reliable and productive methods for exploring the biological aspects of our existence (e.g., physics, chemistry, biology, neuroscience, cognitive psychology). At other times, we are more interested in how our cultural values, practices, and institutions shape who we are and how we think and behave, and there are equally rich traditions of inquiry for exploring our cultural dimensions (e.g., social psychology, cognitive neuroscience, sociology, anthropology, economics, philosophy, history). In short, we need *all* of these methods and modes of explanation, if we want an adequate understanding of our nature as moral creatures.

One of the chief challenges for any naturalistic account of morality is to preserve the complex biological-cultural matrix of relations that make us who we are, and to avoid the temptation of reductionist analysis that treats the biological as ontologically separable from the cultural and as capable of telling the whole story without reference to culture. That said, there are aspects of our biological organism that have little or nothing to do with the fact that we engage others in communities of meaning, value, and practice, just as there are aspects of our cultural engagement that have little or no direct dependence on our physiological makeup.

Consequently, in what follows, "natural" is not intended as a contrast term with "cultural," but rather as a contrast with "supernatural." The only forms of explanation I am rejecting outright are those that posit a realm of transcendent values alleged to exist beyond the world of our embodied, interpersonal, and cultural interactions. My primary reason for rejecting supernatural accounts is, as I will argue, that they do not *explain* anything. Instead, they are merely assertions of a faith in the reality of a transcendent world that is supposed to govern every aspect of our natural world and our lives, but of which we can have no description, no knowledge, and no explanation.

I have no illusions about convincing those who insist on moral absolutes and moral certainty grounded in a supernatural reality, but I shall argue that the human mind simply does not have access to moral absolutes in any cognitively or practically useful sense. I will argue that, in spite of our commonsense belief in absolute foundations of value, we were never, in the history of humankind, in possession of any absolute moral standards, and that we have been deluded in thinking otherwise. In fact, the moral fundamentalist belief in moral absolutes is a recipe for moral obtuseness and avoidance of genuine moral inquiry, and it is therefore an enemy of morality.

Obviously, the view I develop here is not going to be a morality of strict rules, clear decision procedures, unambiguous definitions, or hierarchically ranked moral goods. Nonetheless, it will be able to supply the possibility of genuine moral understanding and psychologically realistic moral guidance. It will provide this guidance by setting out an account of what intelligent moral deliberation looks like. As we will see, one of the most difficult temptations we have to overcome in moral philosophy is our desire for a moral theory that guides us by giving us ultimate moral values, principles, or catalogues of virtues. I will argue, instead, that what we should expect from a moral theory is a psychologically realistic account of intelligent moral inquiry.[1]

A Little Tale of Moral Confusion

I want to begin with a moral adventure story. It is autobiographical and somewhat personal, but I hope that some of the ethical issues I encountered, and some of the questions about the nature of our moral values and practices I found myself struggling with, are questions that need to be addressed in any appropriately critical reflection on the nature of human moral understanding. My personal route to the rejection of moral fundamentalism had, and continues to have, two basic aspects: (1) profound existential doubts about the adequacy of my culturally inherited absolutist moral framework and (2) arguments based on scientific research into the nature of human cognition, judgment, and motivation.

I was born and raised in the Midwest of the United States of America—indeed, in Kansas, which contains the geodesic center of the country and which prides itself on being the *true* "Heartland" of America.[2] Our midwestern values—being, we supposed, God's values—were fit to be everyone's values, or so we thought. My parents raised me to be a good Lutheran and, they fervently hoped, a good Republican. I failed them on both counts. These "failings" were eventually an opportunity for me to rethink my whole conception of what it means to be human, along with my views about the origin of human moral values.

Good Lutherans—the kind of folks Garrison Keillor both celebrates and affectionately makes fun of on his *Prairie Home Companion* weekly radio program—are at least nominally committed to the following view of human nature: (1) Humans were created by an omnipotent, omniscient, and holy God, on whom they are utterly dependent. (2) Every human is born fallen (originally sinful) and cannot save himself or herself without the grace of God. No one can *earn* salvation by any deeds they might accomplish. Everything is a matter of faith and the inward purity of a good will. (3) Humans are created with the moral obligation to realize God's purposes for our lives (and for his creation), by following his commandments and seeking to purify and discipline our will to do what is morally right. (4) Consequently, a moral life is construed as a journey of purification and self-discipline, in order to realize divine purposes. (5) Probably the best that a fallen, fallible human creature is capable of is to seek out the moral standards given by God (as revealed in holy scripture and made incarnate in the actions of Jesus), and then do one's best to live humbly and faithfully by those standards. Any individual person was bound to make some mistakes, but purity of heart and good intentions could go a long way toward making up for your inability to realize the highest good ordained by God.

Although my particular upbringing happened to be Lutheran, I want to suggest that its core assumption is one it shares with any number of culturally different moral systems. This grounding assumption is that humans are fallible creatures whose highest purpose ought to be to cultivate a strong moral character that manifests certain moral values and lives in accordance with certain absolutely binding moral principles. If you set aside, for the moment, the peculiar metaphysics of Christian notions of sin, redemption, heaven, and hell, you might find that there is much to recommend the general ethical orientation just described. Basically, what is required is for us to treat ourselves and others with proper respect (however that gets defined) and to care for the well-being of your soul and that of others. Be loving. Help others in need. Be steadfast. Maintain your integrity. Do not be arrogant or haughty. Realize that the world does not revolve around you. Live to help make the world more nurturing, more kind, and more harmonious. This is an attractive set of moral ideals, a perspective shared by many moral traditions throughout history and across different cultures, regardless of whether or not they are grounded in a theological perspective.

This Heartland picture of religiously grounded morality served to engender in me a strong sense of moral earnestness and obligation. Humans were supposed to recognize their unique place in creation and to understand how it gave them profound moral responsibilities toward themselves and others. By virtue of our distinctive rationality, we alone among the animals were possessed of free will, which imposed on us the moral responsibility to treat all humans with the respect due them in virtue of their intrinsic freedom and dignity. No doubt, this upbringing explains why in college and graduate school I was immediately attracted to Kantian moral theory, which was basically a rationalized version of Judeo-Christian conceptions of universally binding moral commandments. Kant rejected what he regarded as the heteronomous character of most theological ethics, since it placed us under constraints given by an *other* (namely, by God). He replaced the heteronomy of God-given moral commandments with the idea of positive freedom as autonomy (i.e., the giving of moral law by ourselves to ourselves, as an activity of practical reason). Only such self-legislation, he argued, could constitute genuine human freedom. Despite this important difference concerning the ultimate source of moral legislation, however, Kant nevertheless retained a central component of Judeo-Christian ethics, namely, the grounding of morality on unconditional moral laws. In this sense, Kantian rational morality becomes a "de-theologized" version of traditional Christian moral law theory, insofar as divine reason is replaced in Kant's theory by universal reason. Consequently, even though there are significant differences between the heteronomy of divine commandments and the autonomy of universal reason, both views assume the transcendent source of moral values and principles. Hence, for someone (like me) struggling with the problematic ontological assumptions of Christian theologically based moral systems, Kant's vision of autonomously derived, and universally applicable, moral laws offered a welcome alternative.

Unfortunately, there were problems with this absolutist worldview that would not go away. Even as a naïve teenager, with very unsophisticated powers of critical reflection, I immediately discerned some fairly major difficulties with the conception of moral guidance offered both by the religious tradition I had been brought up in and also by the Kantian non-theological alternative version I was still entertaining. These were not highfalutin metaphysical problems (though I would later recognize some of those too), but straightforward issues about how one was supposed to determine which acts and ways of living were right and which were wrong.

The first big problem concerned whether either my religious moral tradition, or its Kantian surrogate, could give guidance to address the range of actual moral concerns any teenager would routinely encounter. Back then, in high school and college, I had hoped that my religious perspective would give me answers to the profound existential questions that any halfway reflective person would end up asking. These were the standard "meaning of life" questions about our existential condition: Is there a God? If there is, what difference should this make for how I live? Or, if there is no God, then what difference should this make for how I live? What is love, and how can I learn to love (and, selfishly, to find love, to be loved)? Why am I here? What am I supposed to be doing with my life? Who am I, anyway? In short, what's this whole human drama all about?

In addition to these grand issues about the nature of reality and our human quest for meaningful lives, there were very concrete and specific moral concerns. For example, there was the profound and pressing question of the "three zones."[3] You know what I am talking about. Zone 1 was from the neck up. Zone 2 went from the shoulders down to the navel. And then there was zone 3—a zone about which young people showed remarkable ignorance, moral uncertainty, and a nearly manic obsession. Good Heartland Christians were not even supposed to think about, much less talk about, zone 3, which meant that it therefore occupied a large portion of the average teenager's interest. The chief problem concerned what you were (morally) allowed to do in those three zones. This was back in the mid-1960s, before the "Summer of Love" in 1968, at a time when the zones were pretty serious business. It is easy to forget how uptight about sexuality we were back then—women wore girdles, people weren't supposed to talk about sex, *Roe v. Wade* had not yet been enacted and so abortion was illegal, and *Playboy* and the lingerie and undergarment sections of the Sears Roebuck catalogue were prime sources of young men's sexual (mis)understandings and fantasies.

Pretty much everyone thought you could kiss all you wanted—go all out in zone 1, although even then there were often qualms about kissing versus "French" kissing. Some people thought zone 2 was more iffy. But why, I wondered. What's the moral difference between copping a good feel (zone 2 action) and kissing someone (zone 1 action) with one of those kisses that went on for who knows how long and steamed up the car windows and left you with a sore tongue the next morning? What supposedly made "petting" worse than kissing? After all, wasn't kissing just a form of petting anyway, only done with the mouth instead of one's hands? And why was "heavy" petting worse than "light" petting? Supposedly it was because as you went from "light" to "heavy," it brought you closer to the mysterious and forbidden zone 3! You were starting up top with the face, lips, and mouth, and then proceeding *downward* toward the place where you could really get into trouble.[4]

I would later come to understand that our entire conception of the ethics of human sexuality—and of our morality in general—rested on a pervasive and unquestioned dualistic metaphysics of the mind-body split. I had learned, mostly from my pastor's sermons and his catechism class, that humans were split creatures, with a mind/soul and a body. The soul—your true inwardness and moral center—was supposedly your highest, most essential self, as well as the seat of your God-given freedom. It was the source of your distinctive rational capacities and the locus of your free will. Consequently, it was your moral center and the source of conscience. In contrast, the body was a problem to be overcome by a purified, disciplined moral will. The body was the source of feelings, emotions, desires, and "the temptations of the flesh" to which we poor humans were subject. To be "good" was to rise above one's bodily, animal nature in order to realize one's true calling as rational soul. The ideal was to be "pure" of spirit and to do your best to retain this purity in a very soiled world.

The point I want to emphasize is that, when it came down to it, nothing in the theological account of morality that I have just sketched really provided any illumination regarding the very pressing concerns I had about matters sexual. One could dredge up strange Old Testament prohibitions against such mysterious forbidden acts as being with a woman who was "unclean" (whatever *that* meant), or sodomy (whatever *that* meant), but if you wanted some good guidance on the ethics of petting, you were not going to find it in the scriptures.

Kant's moral philosophy, which claimed to specify our basic moral obligations via a system of rationally derived imperatives,[5] did not fare much better. Kant has plenty to say about sex, but it is notoriously difficult to justify any of what he says as coming directly from some allegedly pure practical reason. His pronouncements are those of a typical northern European Christian male of his time and place (eighteenth-century Königsburg), and they do not really seem to be the dictates of an allegedly pure practical reason possessed by all rational creatures. For example, as a good German Protestant of his day, Kant claimed that one should not masturbate (which he called "wanton self-abuse"), one should not use another person for sexual gratification "like a lemon to be sucked dry and cast away," and one should not have sex outside marriage. The only way to legitimize sex, according to Kant, was within the context of monogamous marriage. He "reasoned" that in sex you give yourself away to the other as an object to be used by them (which is morally impermissible), and only through marriage could you win yourself back, when they give themselves to you in return![6] In other words, the only way to keep yourself from being reduced to a mere sexual object was to buy yourself back through the marriage contract—you give yourself to them and they give yourself back to you by giving themselves to you. However much these and his other conservative views fit some traditional conceptions of the nature and purpose of human sexuality, they are certainly not issuances of some allegedly universal "pure" practical reason.

In short, at the practical level of day-to-day ethical engagement, neither Judeo-Christian nor Kantian moral law theory provided any serious argument-supported moral guidance other than of the most abstract and vague sort. On occasions when a theory pretended to offer more specific imperatives, I began to notice that there was much *confident assertion* and *little or no compelling argument*. To make matters worse, it seemed to me that none of the other major candidates for systematic moral guidance (e.g., egoism, stoicism, utilitarianism) fared any better when it came to specific moral guidance. Something was rotten in the state of morals, and it was stinking up everything pretty badly, while everyone pretended that our moral understanding was all roses, if only we could smell them.

What was a morally responsible person to do? Neither the Hebrew and Christian scriptures nor Kant's system of morality as set out in his *Metaphysics of Morals* (1797) and his collected *Lectures on Ethics* (1930) could provide even the possibility of a complete systematic response to every actual or conceivable moral problem. Looking for scriptural guidance could never solve your problem, because there could never be full specificity about every conceivable sexual act (or any other types of act, for that matter), and, furthermore, there could never be any algorithm for the proper interpretation of alleged moral commandments.

Consider the problem of the specificity of moral principles. It is not enough for us to know that we ought to love and respect others and ourselves. We also want to know what this means in terms of the important details of our relations with others. We want to know whether it is okay to masturbate, okay to engage in petting to our heart's content, and okay to have sex outside marriage. Is there any place in the Old or New Testament where it says: "Don't engage in petting?" Is there any place where sexual intercourse before marriage is forbidden? The answer is far from clear. Those who believe that premarital sex is immoral will cite passages such as Acts 15:20; 1 Corinthians 7:2, 6:13, 6:18; 2 Corinthians 12:21; Ephesians 5:3; Thessalonians 4:3; Hebrews 13:4; Jude 7; and many others. However, these and other prohibitions against "sexual immorality" give no definition of that term, and it is often suggested by the context that the prohibition in question pertains primarily to adultery and various alleged (but seldom enumerated) sexual "perversions." However, even if there were to be such a scripture, the key question is not really whether there is a passage specifically prohibiting or permitting petting (or sex outside marriage), but, rather, how one was supposed to know what constituted "petting" and what "sex outside marriage" meant.[7] That is to say, how is one supposed to translate pristine moral laws of broad scope into the particular practices of a messy everyday life? Which considerations were relevant to making a correct application of such a rule to the particular cases in hand? Is petting a zone 1, zone 2, or zone 3 affair, and *why* would it be thought to be proscribed for any of these three zones? Is it all right to put your hands on the face and neck of someone you are kissing, but not anywhere else below their neck? Why? And where?

Moreover, what justification could there be for prohibiting certain kinds of intimate relations, short of harming one's partner physically or emotionally? If any kind of sexual prohibition is not just an absolute command of God (or the dictate of some other absolute moral authority), then it requires a justification, and that necessitates some justificatory framework that presupposes an entire moral worldview. However, we would then need a justification for our preferred moral framework, including justifications for its account of human nature, will, agency, emotions, action structure, and so forth. Arguments about justificatory frameworks of this sort were, as I recall, the topics of much heated late-night debate in my college dorm, focused on questions like "What is the allegedly 'natural' end or purpose of sex?" "What kinds of evidence were supposed to determine the 'naturalness' of a practice?" "Why should 'naturalness' even count as a moral value in the first place?" and "Are there any other relevant considerations besides natural teleology?"

As a young man, I thought all of this ambiguity and confusion was probably due only to my personal failings as a moral thinker. I suspected that it was just *my* confusion and *my* inability to think in the right way, the way that would lift the veil of ignorance and reveal to me with blinding clarity the binding moral laws that told me what was permitted and what was prohibited, and exactly under what conditions. However, I soon came to realize that the same types of issues then plaguing me were not just *my* confusions, but instead represented deep philosophical issues that any conscientious and reflective person would have to address concerning the nature of morality itself.

When I pressed others to justify their moral claims, regardless of the basis (theological, rational, cultural, or scientific) they cited for their views, it did not take much careful observation to see that nobody else really had the answers either. I began to notice a common recurring pattern. Whenever someone espoused alleged moral truths with great outward confidence, they typically had almost no idea how to justify their view in a non-circular way.[8] Proclaiming absolute moral truths and manifesting an apparently unshakable confidence in those truths did not equate with actually having any plausible justification, or even any serious understanding of the views they so vociferously espoused. "The Bible tells us so" or "This is what God commands of us" (even if we could agree about what that biblical passage means) is not a rational defense, but only a faithful affirmation of the authority and clarity of a text that itself requires interpretation by fallible human beings. Determining what any set of holy scriptures entails turns out to be a nightmarish hermeneutical process of constructing plausible interpretations of key terms and phrases that are often dramatically underspecified in the text and that may have arisen in a historical or cultural context very different from our own current situation.

INTRODUCTION

Sex was a big enough problem, but things got worse for me when it came to the Vietnam War. In *Moral Imagination* (1993), I have already described my personal dilemma concerning the war. The basic issue was what in the world was a conscientious person supposed to do about participation in that war? I felt like Sergeant York, who believed in his heart "Thou shalt not kill" and yet felt the conflicting duty to serve his country, which was at war. Like Sergeant York, you might quote scripture to justify serving your country in the armed forces, rendering unto Caesar what is due to Caesar, and unto God what is due to God (Matthew 22:21). Or you might also be moved in the contrary direction by the gospel narrative of Jesus, who reputedly said, "You have heard that it was said, 'An eye for an eye, and a tooth for a tooth,' but I say unto you, Do not resist the one who is evil. But if anyone slaps you on the right cheek, turn to him the other also" (Matthew 5:38–39). Then there was "Love one another, as I have loved you" (John 15:12). Could you love your fellow man and go to war with them and kill them? Was there *one right answer* here?

I remember thinking, even as an adolescent, how very convenient it was in the film version of *Sergeant York* that when the young York (played by Gary Cooper) went up on the mountain to pray for God's guidance, a wind came up to blow open his Bible to the page that contained the "render unto Caesar" passage. That divine event apparently answered the question for Gary Cooper's York, but it still left *me* unsure. Sure, I thought, render unto Caesar what is Caesar's, and unto God what is God's, but what is due to Caesar and what to God? How did *that* passage solve anything? On what grounds do we answer the question of *what* is due *to whom*?

Intrigued by the pacifist teachings of Jesus, I was moved to read Tolstoy's impassioned theologically based argument for pacifism in *The Kingdom of God Is within You*, which Gandhi reported as having a profound effect on his own view of nonviolent resistance to evil. However, I got a completely different perspective when I talked with my pastor about the morality of war. He assured me that sometimes horrible things had to be done to realize the ultimate good. He illustrated this with his own personal story about his son, who was a pilot in the U.S. Air Force and was ordered to bomb churches in North Vietnam in which the enemy had sought refuge. Well, maybe, I thought, but on the basis of what theological or ethical principle was it okay to bomb churches? Many pacifists argued that some actions were simply radically incompatible with certain ends, values, and moral obligations. For example, you cannot express

THE NEED FOR ETHICAL NATURALISM

God's love, or even Kantian respect for humanity, by practicing violence on others.[9]

Perhaps you will write off these troubles as the confused musings of a callow youth, which, no doubt, they were. There is no question that at that time I was a typical teenager—naïve, parochial in my experience and vision, and mightily confused about love, life, liberty, and the pursuit of happiness. However, I submit that the kinds of problems I was facing were not just my own personal, idiosyncratic demons, but rather represent basic issues attending *any* moral doctrine or theory that pretends to give ethical guidance by means of unconditional principles, laws, commandments, or standards of value.

For me, it was not until I was in graduate school in the early 1970s that I began to get a glimpse of the true depth of the problems I found circulating in the moral philosophies, both theological and secular, of the day. Back then, it had not yet dawned on me that the solution could not consist in some minor tinkering and refurbishing of inherited views, but instead required a radical reconceiving of the nature of moral cognition and even the point and purpose of moral theory in general.

Given my Lutheran upbringing, with its conception of God as moral lawgiver, it should be no surprise that, as I have already mentioned, when I went off to the University of Kansas in the late 1960s, I gravitated early on toward Kantian moral theory that posited the existence of universally and unconditionally binding moral laws. It seemed right to me that the only way to avoid ethical relativism was a commitment to universal moral laws, or at least universal moral values (e.g., respect), derivable from pure practical reason, and binding on all rational creatures. However, I was already a lapsing Kantian who had emerging doubts about the existence of the kinds of ultimate foundations Kant claimed for his views of knowledge and values. When I then entered graduate school at the University of Chicago, I, like nearly everyone else in philosophy at the time, became enamored of Rawls's quasi-Kantian theory of justice and its implications for the nature of moral theory. If it can be said that Kant attempted to *de-theologize* Judeo-Christian ethics, then Rawls attempted to *de-transcendentalize* Kant, removing Kant's claims of absolute foundations in pure practical reason, while yet keeping most of the rest of Kant's moral vision.[10] So, for a while, I lived on Rawls's Kant-lite diet, which amounted to fundamental rational principles of justice (and concomitantly, I believed, of morality) without any absolute foundation on which to ground them. This is what I mean in saying that Rawls de-transcendentalized Kant by removing what Kant had called "pure practical reason" as a source of moral laws. However, the damage to moral absolutism was done, and I could never go back to any form of moral fundamentalism. Rawls taught us to question foundational, fundamentalist, and absolutist views of moral values, and he taught us that we must not take our moral intuitions as unquestionable. Once those arguments became clear to me, I was utterly convinced—especially after reading Thomas Kuhn, W. V. O. Quine, a number of philosophers of science, John Dewey, and (sometime later) empirical research in cognitive science—that the human quest for certainty, for transcendent foundations, and for pure, non-empirical sources of truth and value was incapable of success, given the limitations of human understanding. Moreover, this was not merely a failed project, but, even worse, adherence to its conception of morality does harm by leading us away from a more humanly appropriate naturalistic conception that is better suited for our lives.

In light of all this, I had to ask myself: "Where does morality come from, if there are no absolute, supernatural, non-empirical foundations for it?" The answer, obviously, is that we must look for natural sources of morality. I will try to work out in subsequent chapters a fairly comprehensive naturalistic ethics orientation. As I conceive it, ethical naturalism starts with the assumption that human moral agents are human animals whose values emerge in ongoing interactions with their physical, interpersonal, and cultural environments. Our values do not come from a noumenal world. Our values are not the dictates of a pure practical reason, because there is no such thing. Perhaps most shockingly, ethical naturalism sees ethical reasoning as a form of problem-solving, and so as not being different in kind from other kinds of hypothetical reasoning and problem-solving that constitute the fabric of our daily lives. Moral deliberation is a process for transforming a morally problematic situation in a way that harmonizes competing ends and commitments.

The obvious problem for any such naturalistic, non-absolutist conception of morality is how to avoid the kind of moral relativism that sees right and wrong, good and bad as entirely a construction of cultural systems. Relativism of this sort entails that there is no point in trying to find a transcendent critical standpoint for assessing the merits of a particular moral system or tradition, since there is no external comprehensive viewpoint from which to judge a particular tradition. Although experimental research on mind, thought, and values leads me to reject any form of foundationalism, I will argue that some criticism of moral ideals and principles is possible, because we are not just prisoners of our inherited moral frameworks. Moreover, it is possible to speak of a particular deliberative process as more or less reasonable than some alternative process, even though we have no God's-eye perspective. Reasonableness will not be a matter of correspondence with some allegedly preexisting universal rationality, but rather will amount to determining how well a certain deliberative process contributes to actually harmonizing previously competing ends and values, resolving tensions, and promoting cooperative, constructive human activities. There is no way of avoiding a plurality of reasonable moral systems and practices, so we should instead focus our attention on how a situated and fallible critical perspective would allow us to engage in reasonable moral appraisal. Like Nussbaum (2000), Hinde (2002), Flanagan (2007), Churchland (2011), McCauley (2011), and many others, I believe that it is possible to say some general things about human nature, without falling into a rigid essentialism. A modest notion of human nature can involve claims about human motivation, sensory-motor capacities, cognitive processes, emotions, needs, social skills, cultural values, and much more. Upon such a minimal conception of human nature, one can give a general account of human well-being and of the best form of moral deliberation for creatures like us. My project is therefore not limited to merely describing values, institutions, and practices, but includes a critical normative dimension.

Two Opposing Conceptions of Moral Value: Non-Naturalistic vs. Naturalistic Approaches

The narrative I have been recounting of my personal struggle to understand the sources and nature of moral guidance is, at its core, a tale of why I came to reject my early non-naturalist views for a more naturalistic perspective. In order to clarify what is at stake in the choice between these two opposed views, I want to focus in a bit more detail on what each of these two fundamental orientations entails.

Non-naturalistic theories locate the source of moral norms and principles in some reality that supposedly transcends the natural world. Those absolute values and principles are believed to be brought to experience, not derived from experience, to give us a basis for assessing the morality of particular actions, moral principles, traits of character, and institutions.

Naturalistic theories, in contrast, see moral values and standards as arising out of our experience in the natural world, which involves biological, interpersonal (social), and cultural dimensions. There is no "pure" *a priori* grounding for moral norms, so they have to emerge from our fundamental needs for survival, individual and group harmony, personal and communal flourishing, and consummation of human meaning and purpose.

Non-Naturalistic Moral Theories

One useful way to understand what is at stake in the contest between non-naturalistic and naturalistic orientations can be fleshed out as a fundamental difference concerning the source and nature of moral guidance. In *Moral Imagination* (1993), I argued that non-naturalistic views are misguidedly obsessed with unconditional moral constraint and governance of our attitudes and behaviors. They therefore tend to take the form of what I called the "Moral Law folk theory of morality" (Johnson 1993, chap. 1). This is essentially the view that the primary purpose of a moral theory should be to provide governance for our actions by specifying rules for which acts are permissible, which are impermissible, and which are morally obligatory. Here, in brief, is that governance theory:

THE MORAL LAW FOLK THEORY

1. Humans have a split nature—a unique conjunction of a mental (or spiritual) dimension and a physical (bodily) dimension.
2. We are driven by our bodily needs and desires to seek satisfactions and pleasures. Because our passions and desires are not intrinsically rational, there arises in all humans a fundamental moral tension between our higher (rational) selves and our lower (bodily) selves.
3. The problem of morality arises only for beings like us who are possessed of a faculty of free will (which is part of our "higher" self), by virtue of which we override, when necessary, our bodily impulses and can control our actions and thus be held responsible for them.
4. Moral constraint comes from a set of universally binding, literal moral principles supplied by revelation, universal human reason, or some other transcendent (supernatural) source.
5. Morally right conduct is thus a matter of (a) discerning what some moral law or moral principle requires in a specific situation, and (b) having the strength of will to do what the moral principle requires, no matter what temptations or influences might make us disinclined to obey the law.

The real core of the Moral Law folk theory is the idea that morality is primarily a set of transcendently grounded universal moral principles or moral properties (values) as the only basis for a non-relativistic morality. Different versions of non-naturalism will identify the source of normative constraint differently, but the three most prominent sources are God, universal reason, and a transcendent realm of non-natural properties. To see what is at stake in non-naturalistic views of morality, let us consider very briefly these three major versions that have exerted considerable influence over moral philosophy in the past two centuries.

1. Classical Theological (Non-Naturalist) Ethics

- God (as the omnipotent, omniscient, and holy creator of all that is) issues absolute moral commands that are binding on all rational creatures (i.e., creatures with reason and free will). Here there are two versions: (1) those in which the commands are simply absolute divine dictates, and (2) those in which the commands are *rational* issuances of the divine mind and so are open to rational reflection.
- Moral reasoning consists in discerning what these unconditional moral laws are and applying them correctly to concrete situations.
- The seat and origin of moral law is thus supernatural (here, divine), and so morality cannot be derived from empirical knowledge of our human capacities, traditions, motivations, or values. On the contrary, anything in human experience must be judged by these transcendent values or principles.

I do not mean to deny that there might be a religious ethics that does not rest on notions of the supernatural. Indeed, I think there are such views (e.g., some types of Buddhism), but they are naturalistic, rather than non-naturalistic, perspectives.

The idea of morality as a system of commandments or moral laws issued by God is the ultimate conversation stopper. If some action is morally obligatory simply because God or Yahweh or Allah or Ramses II commands it—period—then that is the end of the matter, and no rational discussion is necessary or even possible. In moral philosophy, reliance on a notion of the Sacred is a prime example of this form of thinking. Simply to assert that some person, quality, or state of affairs is *sacred* is to attribute to it a supposedly unique and utterly unquestionable moral status that makes it immune to challenge or criticism. Claims such as "Humans (as created in God's image) are sacred" put morality beyond discussion, criticism, and serious reflection. To say that something is sacred effectively insulates it from appraisal by our ordinary means of assessment and judgment.[11] The possibility for rational appraisal and critique arises only if we can get beyond our relation to the Holy or the Sacred as an unquestionable foundational moral reality and ask, instead, what it is about something's being "sacred" that generates moral constraint and responsibility. The possibility of questioning an allegedly revealed

INTRODUCTION

commandment of the Lord God Almighty assumes that God might, in fact, *not* be the ultimate source of moral obligation. It would amount to claiming, as Socrates suggests in Plato's *Euthyphro*, that there is some independent moral standard that even the gods (or God) cannot override. In Western philosophy, this is the basic move that begins a transition from theologically to non-theologically based moral systems. The key to such a transition rests on moving beyond the idea that certain imperatives are unconditionally binding *because* they are commanded by God, to the idea that God's commandments are issuances of a divine *reason* to which humans have some access (through human reason). Only when the commandments or moral laws gain the status of rational principles do they become subject to rational reflection. This is what happens when we move from traditional Judeo-Christian moral law theory to Kantian notions of moral imperatives derived from reason.

2. Kantian Universal Moral Law Theory

Immanuel Kant's moral philosophy is the supreme exemplar in Western philosophy of rationalist non-naturalistic ethics. Kant assumed that the spirit and precepts of his inherited Judeo-Christian tradition were in large measure correct, but instead of locating the source of moral principles in God's commandments (or even in divine reason), he regarded morality as a system of absolute moral laws given by pure practical reason. In other words, as the ultimate basis for moral imperatives, he replaces divine mind or divine reason with universal reason, which is possessed by all rational creatures. With that substitution, pretty much everything else from traditional Judeo-Christian morality can be carried forward without significant change. In both Christian and Kantian ethics, the seat and origin of moral principles remains transcendent and is not derivable from experience. Kant thus argues that empirical considerations are not relevant to the status and nature of our fundamental moral principles.[12] It follows from this that, however much the natural sciences might prove useful in framing a moral psychology or moral anthropology, Kant regarded these empirical disciplines as attendant and utterly subordinate to proper moral theory, which must "investigate the possibility of a categorical imperative entirely a priori, inasmuch as we do not here have the advantage of having its reality given in experience" (Kant 1785/1983, 420).[13]

I will subsequently have more to say about Kantian moral theory, but for now the key point is that the universal reason that is supposedly the source of unconditional moral laws is not thought by Kant to be a product of our nature as animals, of our development as organic creatures, or of our enculturation as social beings. The non-naturalism of Kant's view consists in his claim that although humans possess reason, allegedly as part of their noumenal being, reason is not constituted or shaped by our nature as phenomenal (embodied, physical) creatures.

One of the most characteristic tenets of Kant's moral philosophy—a tenet that I shall criticize later—is that moral laws can only be categorical (i.e., unconditional) imperatives, which allegedly claim purity from anything drawn from "mere" experience. It is for this reason that Kant famously insisted that the sciences, or any other forms of empirical inquiry, cannot contribute significantly to our proper moral understanding.

3. G. E. Moore: Non-Naturalism and the Naturalistic Fallacy

Finally, a quick look at G. E. Moore's *Principia Ethica* (1903) is enlightening, because he gives what is surely the strongest form of ethical non-naturalism ever conceived. His entire theory is framed by two assumptions:

1. Ethics studies the Good (and then defines "Right Action" as that which produces Good).
2. "Good" cannot be defined, because "good" is a simple, unanalyzable, non-natural property of certain states of affairs.

Moore's classic statement of this view runs as follows:

'Good,' then, if we mean by it that quality which we assert to belong to a thing, when we say that the thing is good, is incapable of any definition, . . . and in this sense 'good' has no definition because it is simple and has no parts. It is one of those innumerable objects of thought which are themselves incapable of definition, because they are the ultimate terms by reference to which whatever *is* capable of definition must be defined. . . . There is, therefore, no intrinsic difficulty in the contention that 'good' denotes a simple and indefinable quality. (Moore 1903/1968, 9–10)

Any attempt to define "good" commits what Moore christened the "naturalistic fallacy," because *good* is not a natural object or property. Moore notoriously illustrated the notion of a non-natural property with the property *yellow*, which he claimed we do experience yet which cannot be defined in terms of anything else. Likewise, Moore asserted that *good* is a property possessed by certain states of affairs that we can experience but not define in terms of any natural properties.

My only reason for even mentioning Moore's extreme and, I submit, wildly implausible view is that it would be hard to underestimate the deleterious effect that Moore's work had on twentieth-century Anglo-American moral theory. In effect, Moore silenced constructive moral theory and empirical work on morality for over half of the last century. In my opinion, Moore's meta-ethical views represent a low point in twentieth-century moral philosophy, and I find it distressing that his views continue to receive attention even in contemporary moral theory.[14] I will not argue for this here, but I agree with G. J. Warnock (1962) that Moore's view leaves us with a realm of *sui generis* moral qualities that are supposedly indefinable and yet attach themselves to certain states of affairs, even though we cannot give a rational explanation of why those states of affairs manifest goodness. This makes moral debate about the ultimate values pointless. I would also add that if you find Moore's notion of good as a non-natural property that just happens to alight on certain states of affairs completely unhelpful, then I suggest that you should have correspondingly deep doubts about morality as having supernatural origins in the mind of God, because these views amount to the same thing, from the perspective of analysis, critique, and justificatory explanation. They are both, as I have suggested, equally immune to criticism or rational reflection. As Plato taught us long ago, either something is moral because the gods declare it so or else there is some *reason why* it is the moral thing to do, and providing the appropriate reasons requires assumption of some shared framework of justification.

The Role of Science in Ethical Naturalism

My point in briefly summarizing three versions of ethical non-naturalism is to highlight one key feature they all share, namely, the view that our moral values and principles have their seat in something that supposedly transcends the causal processes and constraints of our natural world. Whether that source of morality is God, pure practical reason, or some unfathomable supernatural moral realm of indefinable properties, the core idea is that no empirical methods of scientific investigation could possibly be relevant to our attempts to explain, and ultimately to justify, certain moral values, virtues, and principles. The non-naturalist argument contends that because the sciences investigate only our natural world and its causal processes, scientific knowledge cannot be relevant to normative evaluations. Science supposedly deals with facts and causal explanation, while ethics is about normative principles and judgments.

Consequently, the biggest obstacle to overcome in developing an ethical naturalism is finding a way to get normative force within the processes of our natural world, without predicating those norms as components of an independent, non-natural realm of values. In later chapters, I will outline many of the natural sources of our moral values and explore the extent to which empirical research can inform and support moral justification. I would only note here that this in no way precludes the need for moral reflection and deliberation, but rather it recognizes the central role of methods of empirical inquiry (especially scientific inquiry) in our processes of moral appraisal and our projection of ideals of moral behavior.

Ethical naturalists typically support their claims about the nature of moral cognition with evidence based on scientific research. Therefore, a word about the status of science is in order here, especially since I will later list some types of scientific evidence that would have a bearing on how we understand moral cognition and also engage in it. The first thing to emphasize is that results of scientific research do not in any way constitute foundational, unrevisable, or absolute knowledge (Kuhn 1970; Putnam 1981). Without rehearsing the many arguments emerging from decades of philosophical debate on the nature of scientific knowledge, I would simply note that the entire field known today as the philosophy of science took its start in the 1930s from attempts to find a foundational grounding for the empirical claims of mature sciences (by which is meant physics). Pretty much the whole history of the ongoing debates about the status of scientific knowledge consisted of ever-widening attacks on epistemic foundationalism of any sort. Once philosophers began to realize that data are theory-laden (i.e., that what counts as a phenomenon to be explained and what counts as relevant evidence are not given *a priori*, but rather depend on the conceptual system and assumptions underlying the particular theory being "tested"), then certain/absolute/presupposition-less/foundational knowledge goes by the board. Instead, we value a scientific theory for the breadth of its evidential support, for its simplicity, for its elegance, for its comprehensiveness, and for many other basic values we happen to regard as important and relevant given our particular context of inquiry. In other words, there is no value-neutral science or scientific perspective, nor is there any way to establish a particular value or set of values as absolutes that are supposedly relevant to any and every theoretical explanation (Rorty 1979, 1982).

The adequacy of any scientific theory or explanatory hypothesis depends on many factors, which cannot be rank-ordered in any absolute or context-independent fashion, and for which there exists no algorithm to specify how to apply the values in question. Moreover, as Bechtel (2009) has shown, most scientific theories that focus on cognition and human psychology do not actually seek universal covering laws (such as, "All X do Y, Z is a case of X, so all Z will do Y"). Instead, most cognitive theories attempt to model non-reductive cognitive mechanisms that would generate the observed behaviors that are of interest to the researcher.

In short, what makes a particular scientific explanation useful is its capacity to generate experimentally manipulable evidence that converges with evidence obtained through other methods and experimental techniques. Sometimes this involves law-like principles, but often it is a form of mechanistic explanation via cognitive models. Since there is no eternally fixed, context-independent, transcendent knowledge, our best strategy of inquiry is to look for converging evidence generated by different methods. The more explanatory methods we can bring to bear on a puzzling phenomenon, and when those methods point toward the same conclusions, the more our confidence increases that we are on a good explanatory path. Converging evidence available from a variety of methods of inquiry is the only way to check the particular metaphysical and epistemic assumptions on which our preferred method or theory rests. Consequently, although there is no basis for any absolutist talk in the realm of science, we can nevertheless gain confidence whenever our experimental inquiry produces results that mesh with results derived from other methods. It is this relative measure of experimental testing that primarily differentiates scientific research from narratives found in religious myths, fictional stories, and cultural narratives.

Science, then, is neither the first word nor the last word, but it is an extremely important voice (or chorus of voices) in our attempt to understand human moral experience, cognition, and judgment. You do not need to believe in absolute scientific truth (which is an illusion) in order to base arguments on scientific results. Some results are so widely recognized as reliable that it is difficult (though not impossible) to imagine them being overthrown. For example, we have mountains of evidence, acquired from different methods, showing that brains work, in part, via complex neuronal patterns of connectivity and activation, that emotions are bodily responses that operate both via activations of certain neuronal assemblies and also via the transmission in the bloodstream of specific chemical compounds (e.g., hormones), and that much of our cognition occurs more or less automatically and beneath our conscious awareness. All of this, and much more, is supported by evidence available from many sources and methods. You can draw *tentative* conclusions from such evidence, even though you cannot claim certainty or infallibility. However, that is sufficient to give us a relatively stable basis for inferences about

THE NEED FOR ETHICAL NATURALISM

moral matters. For example, if it turns out (as seems to be the case) that there is no single executive center in the brain, then any scientifically respectable notion of willing is going to have to involve multiple interacting brain regions operating in parallel and together constituting the initiation of an action by an agent. Any philosophical theory of will that claims radical freedom, or complete transcendence of causal relations in the natural world, would be highly suspect as unlikely to provide a psychologically valid account of human choice (Flanagan 2002).

My main point, then, is that in order to base arguments on scientific research, we do not need to regard science as absolute knowledge. However, neither can we simply ignore scientific research, just because it remains forever susceptible to criticism and revision in light of new methods and new experimental evidence. There are other, non-scientific, forms of evidence that are also important, including phenomenological descriptions of experience, narrative accounts in novels and stories, historical accounts of human affairs, works of art that give us insight into all things human, and so forth. Nonetheless, it is the opportunities for testing, partial confirmation, occasional falsification, and abductive inference (e.g., inference to the "best" explanation) that makes scientific methodologies so reliable and useful. Obviously, although any marshaling of evidential support is potentially subject to possible counterevidence, we proceed to amass as much evidence as we can, using multiple methods, and we draw conclusions advisedly and cautiously, always mindful of the way our best hypotheses can sometimes break under the weight of anomalies and new directions of inquiry.

Given this fallibilist stance on science—or any other form of knowledge, for that matter—I want to end this chapter by giving a short list of some of the major findings of the cognitive sciences that appear to challenge many of the basic assumptions of non-naturalism and that suggest a view of moral thinking that is compatible with our nature as human animals inhabiting our physical and cultural environments. In other words, I want to suggest why ethical non-naturalism is *not* a suitable morality for humans, insofar as it is predicated on a profoundly mistaken conception of human capacities, modes of thinking, and processes of evaluation. What follows is a brief list of the key concerns that seem to me to challenge some of the fundamental assumptions underlying non-naturalistic ethics and the Moral Law folk theory that is often a key part of such theories. In subsequent chapters, I will be delving more deeply into the empirical evidence for many of these claims, so for the present I have provided only a summary statement of them, as a way of showing

INTRODUCTION

why we need to focus on developing a more psychologically adequate naturalistic account of moral cognition. Here, in short, are seven reasons for turning naturalistic.

1. *There is no such thing as a faculty of pure practical reason!* There are several reasons for asserting this, but I will mention just three. First, there is no single place in the brain—no cortical or subcortical area—where everything that goes into our reasoning is done. The brain is a massively parallel functional system, and Enlightenment faculty psychology, upon which the Moral Law folk theory is based, does not fit our current understanding of the brain. Second, even if one granted a so-called "rational capacity," it would not be a pure, disembodied reason. There are a growing number of studies on cognition that reveal the essential role of sensory-motor areas in our capacity for abstract conceptualization and reasoning (Lakoff and Johnson 1999; Barsalou 1999; Damasio 1994, 1999, 2003, 2010; Gibbs 2006; Edelman 2004; Feldman 2006; Johnson 2007; Tucker 2007). In short, the mind is embodied, and we ought rather to speak, as Dewey (1925/1981) did, of the "body-mind," to capture the inextricable process of organism-environment interaction that constitutes our self. "Mind" and "body" are not two different, independent metaphysical entities or dimensions; rather, they are abstractions we make, for various purposes, from the embodied flow of human engagement with the environment. The third reason for giving up the "pure reason" dogma is that human reasoning is inextricably tied up with emotions and feelings, which lie at the heart of our self-monitoring and our values. There is a rapidly mushrooming body of research showing that emotions play a crucial role in social and practical reasoning (Damasio 1994, 1999, 2003, 2010; LeDoux 2002; Prinz 2004; Tucker 2007). Therefore, it is a serious mistake to see morality as the triumph of reason over emotions, feelings, and desires.
2. *Faculty psychology is outdated and misleading.* The Moral Law folk theory appropriates an eighteenth-century faculty psychology that is substantially out of touch with twenty-first-century cognitive science. The old Enlightenment view postulates distinct, independent "faculties" or "powers" of mind (e.g., perception, understanding, reason, imagination, feeling, will) that supposedly combine to generate distinctive kinds of judgments (e.g., scientific, technical, moral, religious, aesthetic, and so forth). Faculty psychology is discredited by contemporary cognitive neuroscience, because we now have extensive evidence that although there are some cognitive "modules," they are of very narrow scope and very specific function (such as motion detectors in the visual system or orientation-sensitive cells), and they do not correlate with any of the large-scale cognitive and affective functions presupposed by traditional faculty psychology (Tucker 2007; Feldman 2006; Bechtel 2009; Edelman 1992; Edelman and Tononi 2000).
3. *There is no single faculty of will.* Obviously, if there are no distinct cognitive faculties, then neither is there a faculty of will. All of the mounting evidence about cognition

and action points to the absence of any single unified center from which action springs. There simply is no single unified self or "I" that is *you*! We certainly do experience ourselves as the unified center of our being and action, but there is no single executive center in the brain that orchestrates our actions (Edelman and Tononi 2000; Bechtel 2009).

4. *There is no such thing as radically free will.* Since there is no single faculty of will, it is not surprising that there is no radically free will. All human freedom is constrained and very limited (Flanagan 2002; Thagard 2010). Freedom of choice is not merely about our so-called "inner resolve," but includes the objective conditions of our current situation and our past history. Dewey saw that what we call our "will" is really a collection of habits of thought and action. The so-called "self" and our "character" are simply an "interpenetration of habits" (Dewey 1922) that define who we are at this very moment. Consequently, willing is less a free action than a result of a complex interpenetration of habits (realized as "weighted" patterns of neural connectivity) that lead to behaviors that we claim as "our own"—that is, as things we have done.[15]

5. *There is no set of universal literal concepts that can be used as elements in universal moral principles.* The Moral Law folk theory specifies the moral "laws" as objective literal principles of right action. The literalist, objectivist theory of meaning that requires that all meaningful concepts be literal is wildly false. Research in the field known as cognitive linguistics over the past twenty-five years has revealed that virtually all our abstract concepts are defined by multiple, often inconsistent sets of conceptual metaphors (Lakoff and Johnson 1980, 1999; Kovecses 2010; Forceville and Urios-Aparisi 2009; Gibbs 1994, 2006; Winter 2001). Not surprisingly, since our moral concepts are irreducibly metaphoric, moral judgment cannot consist in the application of literal moral concepts to real-life situations, as if it were simply a matter of finding the right literal description of a case, in order to see it as falling under a precise literal pre-given principle (Johnson 1993; Winter 2001).

6. *Moreover, the traditional Moral Law folk theory requires the mistaken view that the fundamental moral concepts are determinate and specifiable by necessary and sufficient conditions.* In contrast, years of studies by Eleanor Rosch and many other psychologists, anthropologists, and linguists have shown that many of our concepts exhibit what are known as prototypicality effects (Rosch 1973, 1977; Lakoff 1987). There are central members of a category that fit a central cognitive model, but there are often many non-central members that do not exhibit all the features of the central model (Gibbs 2006; Feldman 2006), and so there is no single univocal definition, even for some of our most important concepts. This means that moral reasoning (like all reasoning) is inescapably open-ended, context-dependent, and relative to the internal structure of our key moral categories, some of which change over time (Winter 2001). The idea that correct moral reasoning consists primarily of bringing a particular case under a more general rule or principle is far too simplistic to

INTRODUCTION

capture the imaginative processes of moral thinking (Johnson 1993; Fesmire 2003). If this is true, then moral literalism and moral fundamentalism are wrong.

7. *Moral absolutism is profoundly mistaken.* However understandable the quest for absolute conceptions of the Good or for absolute, universal moral principles, an overwhelming preponderance of the evidence about human knowing contradicts such a view. One of the great legacies of twentieth-century philosophy will surely prove to be its challenge, from multiple philosophical and scientific perspectives, to any form of foundationalism, absolutism, and value neutrality of human judgments. In Continental philosophy (Lyotard 1979), analytic philosophy (Wittgenstein 1953; Quine 1960; Kuhn 1970), and pragmatist philosophy (Dewey 1925/1981, 1938/1991; Putnam 1981, 1987; Rorty 1979, 1982) alike, there has been a massive sustained challenge to the possibility of presupposition-less, *a priori*, or pure foundations for any form of knowledge. As we will see later, as humans we cannot have this sort of foundational knowledge, nor do we need it to function more or less successfully in our world. We now have several decades of de-transcendentalized, non-foundationalist moral theory to provide suggestions for how to construct a naturalistic view of moral deliberation (Dewey 1922; Rawls 1971; Putnam 2004; Rorty 1989; Flanagan 2007; Hinde 2002).

Most of these seven claims are taken up in considerably more detail in subsequent chapters. For the present, I have but summarized them as unargued assertions only to suggest that if even some of them are substantially correct, then there are deep and persisting problems to be faced about the fundamental inadequacy of several of our mostly taken-for-granted assumptions about morality. The tale I have told about my coming to recognize the need for a naturalistic approach is meant only to recall, for those who know the history of this debate in moral theory, some of the key ideas in the critique of absolutist thinking about morality that need to be taken into consideration. The seven points above are merely some of the more research-oriented (rather than existential) considerations that forced me to abandon the dream of an absolute morality. Unless one swings entirely to the opposite pole of radical subjectivism, relativism, and arbitrariness, then the only other major alternative conception that squares with our best current mind science is some form of ethical naturalism, which sees values and principles as emerging from finite, contingent human experience.

One of the main theses of this book is therefore that the sorts of problems I have just raised about belief in moral guidance via unconditional moral laws are serious, deep, and virtually unavoidable by any moderately reflective person. Moral absolutism is immoral, in that it shuts down precisely the kind of empirically informed ethical inquiry we most need for our lives. Moreover, moral absolutism is dramatically at odds with what we are learning about the nature of moral motivation, thinking, and appraisal. Our best alternative view—one that is supported by a large and growing body of scientific research—reveals that moral deliberation is a process of interwoven imagination, emotion, and reasoning. The bulk of my account will attempt an empirically responsible explanation of the nature, resources, and limits of our moral thinking. The emerging replacement conception of moral deliberation has significant implications for moral guidance, not by way of mistaken views about unconditional moral principles, values, or virtues, but rather in terms of the type of conscientious attitude toward life that gives us our best hope for dealing intelligently with the problems we encounter in our daily lives. Therefore, the account I will be developing is normative primarily in the way it illuminates what it means to engage in proper moral deliberation and the kinds of dispositions we ought to nurture in order to address the types of moral problems we face.

Such a naturalistic view is not only possible but has already for the most part been developed by a number of contemporary ethical naturalists, so we can now indicate the general outline of what such a moral philosophy would involve.

ONE

Moral Problem-Solving as an Empirical Inquiry

A naturalized ethics has to find moral values in our ordinary physical, interpersonal, and cultural experience, and therefore *not* in some alleged realm of pure moral norms and principles. One of the biggest obstacles such an ethics must overcome is the widespread cultural assumption that moral reasoning is a totally unique form of judgment, unlike our ordinary processes of problem-solving in daily life. Morality is thought to be special, because it supposedly imposes unconditional values and universally binding principles, whereas mundane problem-solving is merely means-ends reasoning about how best to achieve a given end or desired state of affairs. Consequently, the first task of a naturalized ethics must be to validate the idea of moral reasoning as a form of problem-solving that does not rely on some special, unique realm of moral values.

Why Should Anyone Think that Moral Problem-Solving Requires a Unique Method?

Why is it that the forms of problem-solving that we employ for the ordinary practical situations we face in daily life are thought by many people to be inappropriate and inadequate when we are dealing with *moral* problems? Why should anyone think that there is some special status for what are known as "moral" or "ethical" judgments that make them fundamentally different in kind from all other types of normative judgments we make in our day-to-day affairs? Why do "moral" problems seem to require for their solution a special kind of method, a special form of reasoning and judgment, a special type of normativity?

One reason that moral strictures are often attributed a unique status is that we tend to believe that moral imperatives trump all our other garden-variety normative judgments, such as appraisals of artworks, claims about the best way to grow good tomatoes, and arguments over what constitutes excellence in medical practice. For example, if your aesthetic preference for veal parmesan conflicts with a perceived moral imperative against the slaughter and consuming of sentient creatures, then your moral obligations are supposed to outweigh any merely aesthetic or prudential values.

I will argue that the existential importance or *gravitas* of moral considerations—which I do not deny—does not require a unique source of values, a unique kind of experience, or a unique form of judgment. In other words, we can certainly acknowledge the importance and special force of moral imperatives without needing to ground that force on a presumption of a moral faculty that generates distinctly moral imperatives. Nor do we need some special exalted source—beyond our ordinary experience—to generate the moral obligation we experience.

I suspect that what leads many philosophers to think that moral reasoning is a unique and utterly distinct form of thinking is their mistaken belief that moral questions arise from a unique and utterly distinct *type* of experience—moral experience (as opposed to, say, aesthetic, religious, political, or economic experience). Moral questions, they reason, are questions about norms that generate imperatives about what unconditionally *ought* to be done—about right action. And the "ought" in such cases is supposed to have a certain overriding force. Moral norms pertain to moral rights, responsibilities, and actions, and they are not merely technical norms specifying what ought to be done to achieve a certain effect or desired state (as in "you ought to fertilize with chicken manure if you want the best flavor in Big Boy tomatoes"). Consequently, these moral norms and principles are thought to have a special normative force that is not present in other types of experiences and other forms of judgment. Those norms, they claim, must come from some pure source: either divine mind, an alleged universal pure practical reason, or perhaps some other transcendent origin—a source that generates commands of reason concerning which actions are morally permissible, which are impermissible, and which are obligatory.

CHAPTER ONE

I do not see any need to posit a special transcendent source of normativity to ground the force of moral judgments. Whatever moral "trumping" force there might be is primarily a consequence of the fact that certain things tend to matter more for us because they are thought to be necessary for the well-being of ourselves and others. My contention is that moral deliberation is a process of problem-solving that arises in situations of moral indeterminacy and conflict. I do not see any compelling evidence that we need a special form of normative reasoning that we do not already employ in dealing with other problems in our ordinary experience where values are at stake. We do not need some special moral faculty designed to deal exclusively with moral problems. Instead, moral inquiry involves valuations and appraisals that are similar in kind to judgments about good farming techniques, exemplary medical practices, or aesthetically compelling works. It is past time to get over the mistaken idea that so-called "factual" claims have neither a normative dimension nor normative implications. In this chapter, then, I will criticize the notion that I call the "autonomy (self-legislation) of the morally normative" and the attendant idea that there are distinctly *moral* experiences or problems that require a distinctly *moral* form of reasoning or judgment, based on distinctly *moral* values.

Toward the end of *Human Nature and Conduct* (1922), John Dewey says that his "foremost conclusion is that morals has to do with all activity into which alternative possibilities enter. For wherever they enter a difference between better and worse arises. Reflection upon action means uncertainty and consequent need of decision as to which course is better" (193). Moral deliberation arises when you are confronted with a problematic situation that calls for a decision about which of the perceived competing possibilities for action are better and which are worse. Dewey famously describes this moral problem-solving as a deliberative process of imaginative exploration of available alternative courses of action:

> We begin with a summary assertion that deliberation is a dramatic rehearsal (in imagination) of various competing possible lines of action. It starts from the blocking of efficient overt action, due to that conflict of prior habit and newly released impulse to which reference has been made. Then each habit, each impulse, involved in the temporary suspension of overt action takes its turn in being tried out. Deliberation is an experiment in finding out what the various lines of possible action are really like. (132)

The view represented in these two statements strikes me as so obvious that it is difficult to imagine anyone taking issue with it. Nevertheless, some version of it has been summarily and stridently rejected again and again over the past two centuries of moral theory. I will later develop my Deweyan view of imaginative moral deliberation, but for now I want to explain and defend the claim that moral reasoning is a form of problem-solving activity that utilizes basic forms of normative judgments of the sort that we employ all the time in our day-to-day experience. I do not imagine that anyone seriously denies that we engage in moral problem-solving, because you would not even need deliberation unless there was at least some *prima facie* problem facing you about how to act. Therefore, the issue is not whether moral reasoning is a form of problem-solving, but rather *what forms of inquiry* are appropriate to moral problems. The view I want to challenge is the idea that moral judgments require a unique source of values beyond what is involved in our ordinary problem-solving activities and a unique method of reasoning that is not a component of other basic kinds of inquiry.

Let us begin by considering a very mundane form of problem-solving. Let's say you want to grow luscious tomatoes for your dinner table. You get advice from experienced, knowledgeable farmers about how to prepare the soil, what varieties of tomatoes grow best under which climate conditions, how to plant the starts, how to tend them, how much and what kind of fertilizer works best, what to do about insects and fungi, how much watering is recommended, and when to harvest the fruits of your labors. Behind all of this expert knowledge is years, perhaps even centuries, of accumulated scientific (especially botanical) knowledge coupled with the know-how of experienced practitioners (farmers).

Farming is what Alasdair MacIntyre calls a "practice," by which he means "any coherent and complex form of socially established cooperative human activity through which goods internal to that form of activity are realized in the course of trying to achieve those standards of excellence which are appropriate to, and partially definitive of, that form of activity, with the result that human powers to achieve excellence, and human conceptions of the ends and goods involved, are systematically extended" (1984, 187). The appropriate ends and goals of farming are established by a long history of expert practitioners (i.e., farmers) who come to understand what good farming is and what it can achieve when it is done well. Given the goods or ends defined internal to the practice of farming, farmers learn to recognize and prize exemplary enactments of their practice. This, in turn, gives rise to catalogues of the virtues that are required to realize the ends appropriate to that particular practice. The virtues (excellences) of good farmers are the traits of body and mind—the physical and mental skills and dispositions—that generally lead to success (i.e., exemplary achievement) in farming.

Practices are dynamic, not static. They can evolve through the creative activities of people who engage in them—that is, through ongoing experimentation with materials and techniques, through the transformation of the ends that define the practice, and through the emergence of extended or wholly new conceptions of ends and excellences. In such a conception of human activity, the values and standards of good practices come from within, and may be modified within, the developing practices. They are not external impositions from outside the practice itself, although there may certainly be "extrinsic" ends, purposes, and values that are manifest across a wide variety of different traditions of practice (e.g., making money, gaining respect, becoming famous, etc.).[1]

Every practice thus develops at least some recognized processes of inquiry and experimentation for carrying out the practice, and even for reconstructing it when the occasion calls for adaptation. In each of these cases, there is clearly some form of engaged inquiry going on, there is utilization of skills appropriate for excellent performance, there is reliance on empirical (often scientific) knowledge relevant to the solution of particular problems, and occasionally there is innovation in creating new modes of activity that extend or redefine the practice.[2]

MacIntyre's notion of practices leads to the idea that living a morally exemplary life requires developing the requisite virtues, employing the appropriate forms of inquiry, and exercising the requisite skills necessary for navigating the sociocultural landscape in which one finds oneself. As Paul Churchland phrases it, moral knowledge is

> a *set of skills*. To begin with, a morally knowledgeable adult has clearly acquired a sophisticated family of *perceptual* or *recognitional* skills, which skills allow him a running comprehension of his own social and moral circumstances, and of the social and moral circumstances of the others in his community. Equally clearly, a morally knowledgeable adult has acquired a complex set of *behavioral* and *manipulational* skills, which skills make possible his successful social and moral interaction with the others in his community. (2007, 40–41)

An important part of the requisite skill for navigating an extremely complex social landscape is the ability to assume a critical and reconstructive perspective on one's own inherited moral tradition, so that we do not simply reproduce its values and practices blindly. Without this skill, there can be no intelligent moral growth and transformation. I address this critical and creative dimension later, in my treatment of moral deliberation, but for now I am interested in the idea that moral cultivation is acquiring virtues, skills, and attitudes that allow one to resolve morally problematic situations.[3]

MacIntyre's account of practices captures the way in which forms of inquiry and problem-solving, with their own standards of excellence, emerge *within* the practices themselves, without requiring guidance from some source external to the practice. The question is: Can the methods and standards of moral inquiry emerge out of traditions of practice within moral communities, or do we need some external source of norms to guide our appraisals and behaviors? I will argue later that there are no such external, transcendent, or supernatural sources of moral values and principles, and that even if they did exist, we would not need them for our moral deliberation.

Given the conception of inquiry as embedded in, and emerging from, complex practices, including the ability for the inquiry to turn back on the practice itself to transform it, we can now return to the question with which I began this chapter: namely, why, when it comes to what we call our "moral" problems, would anyone refuse to utilize the same methods of inquiry and experimentation that we have developed through centuries of laborious practical engagement with problem-solving in our everyday lives? Why should so-called "moral" problems be thought of as essentially different in kind from our other garden-variety practical problems? And especially, why would we need a completely different method for solving those problems and a completely different set of standards or values of success and right action?

The answer to this fundamental question is that we have inherited a mistaken view of ethical questions as *sui generis* and therefore as requiring a unique set of standards, norms, and modes of reasoning that are fundamentally different in kind from our other kinds of mundane problem-solving. This view is an error that arises from our understandable desire for a moral certainty that we think could only come from universal, transcendent, unconditionally binding values and principles that would allow us to assess the relative merits of all moral systems that have ever existed.

My response to this received notion of unique types of experience and their corresponding unique forms of judgment is that (1) we need to abandon the idea that there are distinctive "moral" experiences (or problems) that are radically different in kind from other alleged types of experience (e.g., technical, scientific, aesthetic, religious), and (2) we need to give up the idea that this supposedly unique type of experience requires a corresponding unique type of judgment or reasoning for discerning how we ought to act.

I am going to follow Dewey's lead in claiming that the possibility of a workable naturalized ethics rests on recognizing moral reasoning as a form of problem-solving that ought to employ the best methods of inquiry and the best experimental procedures available. Unfortunately, we are saddled with centuries-old traditions of moral philosophy that regard the idea of empirically grounded problem-solving as anathema to a proper conception of moral reasoning. Most of the arguments against the viability of a naturalistic ethics rely on some version of the claim that there are distinct moral experiences requiring distinct forms of ethical reasoning that are essentially different from the reasoning involved in other types of reasoning about practical matters. So, it behooves us to understand the reasons why the "autonomy of the ethical" is so deeply rooted in our commonsense and philosophical theories alike. To see how this fundamental differentiation of types of situations and types of judgments arose is to discover where certain kinds of moral philosophy went wrong and why ethical naturalism is still today accused of committing G. E. Moore's supposedly catastrophic "naturalistic fallacy."

The Myth of a Uniquely Moral Kind of Experience

Early in his book *Moral Minds: How Nature Designed Our Universal Sense of Right and Wrong* (2006), Marc Hauser tells a fictionalized story of a young girl who is chastised by her father for two allegedly different kinds of actions, one supposedly moral and the other supposedly practical (what Hauser calls "commonsensical"). In the first case, the girl hits a boy who refuses to let her play in the sandbox. In the second case, she puts sand in her mouth. In both cases, her father expresses his anger and strong disapproval of the action. Hauser claims that the girl experiences the same anger and disapproval from her father regarding the two different cases, but somehow she recognizes that the causes of that disapproval represent two fundamentally *different kinds* of acts—one moral and the other merely a matter of hygiene (and hence a practical, commonsense issue). Hauser concludes, "Same emotion, different conclusion. Hitting has moral weight. Eating sand does not. How do the child's emotions send one action to the moral sense and the other to common sense?" (30).[4]

Hauser's example captures the widely held intuition that there are several independent kinds of experience, each requiring its own distinctive correlative form of reasoning and evaluation. The basic idea is that human experience comes differentiated into distinct types—theoretical, technical, moral, aesthetic, political, religious, and so forth—and that it is from our parents, teachers, experts, and peers that we learn how to categorize experiences not usually by explicit instruction, but more often simply through life experience. The notion that there are different discrete kinds of experience is a commonplace in many cultures, and there is at least some evidential support for the view. For example, in Western traditions, we tend to classify some actions and judgments as *ethical* (e.g., abortion, lying, murder, altruism), some as *theoretical* (e.g., the atomic weight of gold, whether Pluto is a planet, whether tamoxifen can help prevent the recurrence of breast cancer), some as *technical* (e.g., how to deactivate a bomb, how to make bread, how to cure a disease), some as *aesthetic* (e.g., whether a symphony manifests more the classical or romantic style, whether a certain sculpture is beautiful, which of Monet's paintings of the same haystack are better), and so on with other supposed types of experiences. So, it begins to look as if our experiences come preclassified into separate kinds, each with its own special character that is not shared by any other kind.

Moreover, there is an impressive body of research into the way that early in their moral development children across cultures make a distinction between acts that are moral in nature and those that are merely social conventions. Elliot Turiel, a principal researcher in this area, claims that "over a period of more than twenty years, nearly 100 studies have been conducted that support the proposition that children make judgments that differ in accord with the moral and conventional domains" (2002, 110). The basic idea emerging from such studies is that early on children learn to distinguish between *moral* strictures governing acts that involve potential harm, both physical and psychological, to oneself and others, and *social conventions* established by some social or group authority. Turiel summarizes the key distinctions as follows:

Conventions are shared behaviors (uniformities, rules) whose meanings are defined by the social systems in which they are embedded. Therefore, the validity of conventions lies in their links to existing social systems. Morality, too, applies to social systems, but contrasts with convention in that it is not determined by existing uniformities. As delineated by moral philosophers, moral prescriptions are not specific to a given society; they are not legitimated by agreement; and they are impartial in the sense that they are not determined by personal preferences or individual inclinations. (110)

CHAPTER ONE

Turiel illustrates this distinction with an interview of a five-year-old boy who is asked about two situations, one in which children in a preschool are allowed to hit each other, and another in which children are permitted to be without clothes on warm days. The child insists that hitting is "not okay," and when asked why not, he responds, "Because that is like making other people unhappy. You can hurt them that way. It hurts other people, hurting is not good" (109). Furthermore, children tend to regard the moral restrictions of this sort to be binding, even if a teacher or other authority tells them that it is permissible to hit other children. In other words, the source of moral obligation appears to transcend any social or conventional authority.

In contrast, the boy recognizes that some constraints are merely socially and conventionally imposed by authorities who are empowered (by communal agreement) to enforce those constraints within the confines of a particular conventional framework. Thus, the five-year-old reasons that it would be okay for children not to wear clothes, if that is what is decided by duly appointed school authorities: "If that's what the boss wants to do, he can do that."

It would seem to follow, then, that moral judgments and prescriptions transcend social customs and practices, potentially overriding any conflicting social conventions. Children learn to judge that certain kinds of harm are simply morally impermissible, even if some individual or group authority claims to override the ethical prescription:

> Children and adolescents judge moral obligations not as contingent on rules or authority and as applicable across social contexts. Moral transgressions, such as hitting or stealing, are not judged by the existence of rules, the directives of authorities, or commonly accepted practices (e.g., the act is wrong even if it were acceptable practice in a culture). Rather, rules pertaining to moral issues are judged as unalterable by agreement, and such acts would be wrong even if there were no rules governing them. (Turiel 2002, 111–12)

I do not question the validity of these studies, the robustness of this type of finding, or even the cross-cultural evidence regarding the appreciation, even among children and adolescents, of a fundamental distinction of moral versus conventional practices and norms. However, I would argue that the widespread cross-cultural existence of such a differentiation between the moral and the conventional need not be taken as justifying claims that such a distinction marks some essential ontological or epistemological difference between experiential types. Some types of actions are understood by us to profoundly affect the well-being of ourselves and others, and these actions typically involve notions of helping, harming, care, justice, and respect (to name but a few dimensions). Other behaviors are rightly seen as only matters of socially sanctioned practices, perhaps non-universal, and subject to being overridden by moral obligations.

I submit that this moral/conventional distinction is a matter of degree, and so will at best establish only a culturally contingent continuum, with clearly moral issues on one pole (e.g., Thou shalt not murder), clearly conventional prescriptions lying at the opposite pole (e.g., Thou shalt not eat peas with your knife; thou shalt not talk with food in your mouth), and a vast range of intermediate cases lying at various points between the two poles (e.g., Thou shalt not enter a church, synagogue, or mosque without a head covering if you are a female; thou shalt respect your elders; thou shalt not wear revealing clothing in public; thou shalt vote in public elections; thou shalt not dance). What children (and adults) are realizing in marking the moral/conventional distinction is that some attitudes and behaviors are more important for our individual and communal well-being, while others are mostly a matter of group preferences relative to social practices within communities. However, there can be no value-neutral or absolute non-question-begging way to draw a rigid distinction between these two types of experience.

Again, I do not wish to deny that we *do* make such classifications. We routinely use expressions like "She's faced with a really hard *ethical* choice," "It was merely a *pragmatic* (or *practical*) decision about how much money to spend on the wedding gift," "It's a simple *technical* question about whether to use oil-base or latex paint on your house," or "It's a matter of *aesthetic* preference whether you prefer van Gogh over Warhol." Qualifying terms such as "moral," "pragmatic," "technical," and "aesthetic" seldom get us into any trouble when they are used principally to mark some dominant quality of a situation in a particular context of action. For example, when the dominant quality of a situation involves the well-being of oneself and other people (including notions of harm, justice, rights, care, etc.), we tend to identify that as a "moral," rather than an aesthetic or a technical problem. A good moral theory will need to explain why some types of considerations are considered by various communities to have a certain unconditional force, while others are regarded as conventional and socially contingent. However, this can be done without recourse to a realm of pure moral norms.

These alleged *kinds* are not intrinsic to the essential nature of certain experiences. Experiences do not come premarked with their "type" stamped on them in indelible metaphysical ink. Rather, these differentiating adjectives like "aesthetic," "moral," "technical," or "matters of

CHAPTER ONE

etiquette" are the product of culturally influenced selections that we make about what is important to us in a given context, and these selections are often relative to what we care most about at a given point in time. They are matters of emphasis, relative to our concerns and interests, rather than being essential ontological structures embedded eternally "in the nature of things." Moreover, the ways we classify various actions may sometimes vary from culture to culture, such as when some action regarded as "merely a matter of aesthetics" in one culture is seen in a different culture as having profound moral implications. For example, the way one dresses in public places might be considered merely a matter of style and aesthetic preference in the United States, but might have profound moral implications in an Islamic culture. Turiel (2002), for instance, cites an account by Fatima Mernissi (1994) of her life as a young girl growing up in a harem in Morocco, in which acts by women, such as chewing gum, listening to the radio, or dancing—all of which were forbidden by men—came to be "revolutionary gesture(s)" and not merely aesthetic or social preferences. Is chewing gum or listening to the radio intrinsically or *essentially* a moral issue, or is their moral weight culturally contingent? There is considerable evidence that different cultures often carve up "moral" considerations differently (Shweder 1984, 1987, 1997; Henrich, Heine, and Norenzayan 2010). As Jonathan Haidt concludes, "The moral domain varies by culture. It is unusually narrow in Western, educated, and individualistic cultures. Sociocentric cultures broaden the moral domain to encompass and regulate more aspects of life" (2012, 26).

Even within a culture, there is often widespread uncertainty about how to classify a particular action. Is your neighbor's decision to park his recreational vehicle or his eighteen-wheeler long-haul truck in his front yard a practical decision, an aesthetic choice, or a moral issue? In American culture in certain types of neighborhoods, it would seem to involve *all of these dimensions and more*. Is the freedom to smoke tobacco a political issue, a moral issue, a health issue, an aesthetic issue, or all of the above? I recently read in my local newspaper of a man who painted his house with polka dots. Apparently, the city had promised to buy his home to secure land for a new development, but they had not followed through on the sale. He stated that perhaps *now* the city might get around to the purchase, and he was asking $50,000 more than the original price. The article concluded with the statement that his neighbors could not have been happy about this action. Was his act an aesthetic, political, economic, or moral act? The answer is: Yes (all of the above).

The fact of the matter is that in most cases a developing experience involves multiple dimensions, which are often inextricably intertwined, so that selecting just one descriptive category misses the richness, complexity, and depth of any given experience. For example, what we tend to call a "health" issue cannot be understood fully without appreciating the economic, prudential, moral, aesthetic, psychological, and social aspects of the experience.[5] Anyone who works in a health-care profession knows full well that any given "health" issue will typically have economic, social, aesthetic, psychological, spiritual, and political dimensions too. In claiming that decisions about types of experiences are matters of emphasis and selective interest, Dewey meant that we select certain qualities or dimensions of a situation as salient in relation to some goal, value, interest, or purpose that stands out as important for us in that situation. In Western cultures, we typically call an experience "ethical" or "moral" to emphasize that issues of the harm or well-being of ourselves and others are a prominent concern at that particular moment. We tend to call an experience "aesthetic" when we are focused primarily on the qualitative dimensions of a situation (say, a sunset, an artwork, a building). We call an experience "political" insofar as we are concerned with civic institutions and practices that affect large communities or social organizations. We call an experience "practical" (or a technical matter) when we are attending primarily to the means to realize either some pre-given end or else some end-in-view that we have for the moment. We call a choice "prudential" when it involves in a central way the happiness of one or more people.

The moral here is that any experience we mark off within the mundane flow of events in our lives and select for attention is likely to have *all* of these dimensions and more interwoven to constitute the experience as lived. Dewey gives an example of three such aspects of experience:

It is not possible to divide in a vital experience the practical, emotional, and intellectual from one another and to set the properties of one over against the characteristics of the others. The emotional phase binds parts together into a single whole; "intellectual" simply names the fact that the experience has meaning; "practical" indicates that the organism is interacting with events and objects which surround it. The most elaborate philosophic or scientific inquiry and the most ambitious industrial or political enterprise has, when its different ingredients constitute an integral experience, esthetic quality. (1934/1987, 55)

In short, any specific experience will be mottled, manifesting a variety of dimensions, with some more prominent or relevant to our particular interests in a given situation than others. There is no such thing as a merely and exclusively "moral" experience. Consequently, as we will see in the next section, it is unlikely that there should be a distinct type of judgment that is alone appropriate for a so-called "moral" experience.

My claim is that even in contexts where we quite appropriately distinguish "the moral" from "the conventional" (or any other qualifying adjective), we need to realize that what we designate as *moral* nearly always incorporates selected dimensions or aspects that might include various mixtures of other factors, such as aesthetic, prudential, technical, emotional, intellectual, religious, political, or economic considerations. In other words, the relatively unproblematic partitioning of kinds of experience that operates in daily life can become problematic if it leads us to oversimplify a situation and thus ignore its complexity. The error is to think that there is some preestablished ontological, epistemic, or phenomenological absoluteness about our preferred distinctions.

Coming to see that certain types of actions are matters of harm to others or oneself is obviously an important insight upon which our individual and communal well-being depends. However, it most assuredly need not follow from this that moral matters transcend cultural systems, as if etched for all time in metaphysical diamond. Instead, what I have been objecting to is the claim of Hauser (2006), and many others throughout history, that there exists a set of exclusively "moral" issues that involves a unique set of moral strictures and requires a unique type of moral judgment process. That, I will argue, is a non sequitur, because the fact that we regard certain issues as predominately moral does not entail that what defines those issues as moral somehow excludes a number of other dimensions of human experience. We expect and desire that children and adults alike learn that certain actions are morally significant and are not merely an issuance of some social authority; but we need not, and should not, infer from this that there are purely "moral" experiences having no admixture of aesthetic, religious, scientific, technical, economic, or political dimensions interwoven into a complex tapestry of a given situation. I shall argue that reducing experiences to exclusive kinds is precisely the sort of reductionism that leads us to overlook or under-appreciate the complexity of a particular problematic situation that requires our reflective reconstruction.

The Myth of a Uniquely Moral Kind of Judgment

The central point I want to emphasize here is that our practices of classifying experiences into discrete kinds may serve various purposes of inquiry, decision-making, and social coordination, but the kinds do not represent essential qualities, absolute ontological distinctions, or some allegedly mind-independent or culture-transcending objective character of a given situation. As long as we are able to keep this point in mind, we can avoid being drawn into the error of believing that there are independent and autonomous kinds of mental judgments appropriate to each alleged type of experience. In our mundane affairs, we will routinely speak of experience as if it came sorted into kinds, and, for the most part, this causes no problems. But there is much mischief that ensues when philosophers and psychologists erect these commonplace categorizations of experience into a theory of distinct kinds of judgments, employing distinct kinds of values, as the basis for the character of these allegedly distinct kinds of experiences.

In investigating how an individual or a group classifies its experience, one comes repeatedly into contact with a problematic general framework that I have elsewhere called "the folk theory of Experiential Kinds":

1. Experiences come differentiated into distinct, discrete, independent kinds (e.g., practical, moral, prudential, technical, intellectual, aesthetic).
2. For each particular kind of experience, there is believed to be a correlative form of judgment appropriate to it (e.g., technical judgments, ethical judgments, prudential judgments, aesthetic judgments, etc.).
3. Each form of judgment is the product of our various cognitive faculties (e.g., perception, imagination, feeling, understanding, reason, will) operating together in specific relations.
4. Each kind of judgment employs its own unique standards and values. For example, it is held that deciding whether a contemplated action is morally permissible is quite different from deciding whether it will make you happy, whether it can be accomplished quickly and easily, or whether it is aesthetically significant.
5. Each of those standards, principles, or values must have its own unique source in the mind (or beyond the mind).

This folk theory of Experiential Kinds has been assumed in virtually every major moral theory that presupposes any form of the faculty psychology and theory of judgment developed in the Enlightenment and carried down to the present day. It is now so deeply interwoven into the fabric of our contemporary ethical perspectives that it has come to seem almost intuitively self-evident to most people, despite the fact that it is an artifact of a particular Enlightenment perspective, generated within the context of a historically contingent set of interests, practices, and values.

Perhaps the most radical, and the most influential, version of this folk theory applied to morals is Kant's celebrated moral philosophy. Kant

CHAPTER ONE

famously framed the fundamental problems and methods of philosophy in a manner that fatefully directed much subsequent philosophy. In his ethics this was especially true, insofar as his particular characterization of moral experience and judgment profoundly influenced the way later moral philosophy thought about the nature of moral reasoning. In brief, what is known as Kant's "Critical Philosophy" gravitates around one central question: Given the alleged character of a certain type of judgment (as appropriate to a certain type or domain of experience), what are the conditions of possibility for, and the limits of, that type of judgment? Kant assumes that the fundamental mental process of thought is what he calls "judgment," in which disparate elements (such as sensations and concepts) are unified into a coherent structure of thought.[6] Each form of judgment rests on specific conditions for its possibility, such as the operations of our mental faculties in various relations and directed toward various types of objects, and it is the task of critical philosophy to explain how those judgment types are possible and what their limits are. So, in the *Critique of Pure Reason* (1781), the judgment type that needs explanation and justification is one that issues in objective knowledge of our world (especially in the form of scientific knowledge). In the *Critique of Practical Reason* (1787), the judgments needing justification are moral, and in the *Critique of Judgment* (1790), the focus is aesthetic judgments and teleological judgments of nature. In short, each type of judgment is alleged to exhibit certain defining characteristics that need to be explained and justified if that judgment is to be valid.

In his *Grounding for the Metaphysics of Morals* (1785), Kant thus takes up the problem of how moral judgments are possible, how they can claim universal validity, and how they can be absolutely binding on humans and all rational creatures, as matters of unconditional duty. One of the surprising features of Kant's moral philosophy is the absence of any serious argument to support his characterization of moral judgments. In the preface to the *Grounding*, for instance, he merely asserts the necessity of a "pure moral philosophy," by which he means one grounded in pure practical reason and not influenced by "anything empirical."[7] And how do we come to this foundational insight? Kant answers:

> That there must be such a [pure moral] philosophy is evident from the common idea of duty and of moral laws. Everyone must admit that if a law is to be morally valid, i.e., is to be valid as a ground of obligation, then it must carry with it absolute necessity. . . . And he must concede that the ground of obligation here must therefore be sought not in the nature of man nor in the circumstances of the world in which man is placed, but must be sought a priori solely in the concepts of pure reason; he must grant that every

other precept which is founded on principles of mere experience—even a precept that may in certain respects be universal—insofar as it rests in the least on empirical grounds—perhaps only in its motive—can indeed be called a practical rule, but never a moral law." (1785/1983, 389)

Now, it would be a nice achievement if you could get everyone to admit what "everyone must admit" and what is "evident" (389), but that is famously impossible, or at least notoriously improbable. What Kant thinks "everyone must admit" is that moral law commands with absolute necessity (unconditionally), and that such necessity cannot be derived merely from an account of experience (which would describe what is, but not what must be). However, if you do not agree that moral law is characterized by necessity, then you will not think it obvious that a morality derived from experience is no morality at all. Kant's reliance on universal assent to his key proposition is predicated on his audience sharing his culture and values, even though he would deny this. It would appear that the "everyone" who "must admit" that Kant is right turns out to be those who share at least most of the key concepts, values, and principles embedded in the Judeo-Christian culture he took for granted. And not only that, for they must also share most of his assumptions about the types and nature of judgments, the nature of our mental faculties, and our modes of knowing.

However, what if you do *not* agree with Kant that (1) moral law commands unconditionally, and with absolute necessity, and that (2) only principles derived *a priori* from pure practical reason (and not drawn from experience) can serve as moral laws? Furthermore, what if, as recent research in cognitive science suggests, you regard moral principles as summaries of practical considerations that a culture has found useful in dealing with its broadly conceived moral problems? What if you deny, on empirical grounds, the existence of an allegedly "pure" (nonempirical) practical reason? What if you reject, based on extensive critical analysis from various philosophical orientations over the last century, Kant's views about the types of judgments and the nature of reason? Then Kant's whole system turns out to be an extremely insightful rational reconstruction of his culturally inherited Judeo-Christian moral framework (with certain theological assumptions removed), but nothing like a transcendently grounded set of absolute moral laws.[8] If you deny (1) above, as I would, then you do not even have to address (2).

Nevertheless, for the sake of seeing where Kant's argument has so fatefully led us over the last two centuries, let us trace out some of the implications of his basic assumptions. Once you accept (1) and (2) above, the

CHAPTER ONE

rest is Kantian history. For, if you assume (a) that there is a uniquely *moral* kind of experience (and corresponding form of moral judgment), (b) that it is governed by principles that hold unconditionally for all rational beings (and not just humans), and (c) that only pure practical reason could possibly generate the unconditional (universal and necessary) character of that type of judgment, then you get Kant's famous absolutist theory of moral law. That theory rests on making a strict separation between two fundamental types of imperatives: "Now all imperatives command either hypothetically or categorically. The former represent the practical necessity of a possible action as a means for attaining something else that one wants (or may possibly want). The categorical imperative would be one which represented an action as objectively necessary in itself, without reference to another end" (1785/1983, 414). Hypothetical (conditional) imperatives are formally characterizable as "If you want [i.e., will] X, then you ought to do Y [as the necessary and indispensable means thereto]." Categorical (unconditional) imperatives amount formally to "You ought to will Y [regardless of any other ends you might have]." The crux of Kant's view is that moral laws, insofar as they are universally and unconditionally valid, can only be categorical imperatives. Otherwise, they would be mere hypothetical imperatives specifying necessary means to pre-given ends, and so they would be merely conditional and would lack the requisite purity of origin that "everyone must admit" is the defining character of moral principles.

On the basis of this false dichotomy of types, Kant then proceeds to reject any notion of moral thinking as hypothetical reasoning geared toward problem-solving. Even worse, he believed that any empirical study of human nature, thought, and motivation would serve principally in framing and realizing hypothetical imperatives, but would in no way determine universal moral laws. Moral principles/laws (as imperatives compatible with the "one and only" categorical imperative) must be derived from pure practical reason, and they cannot be based on anything empirical. Hence, empirical considerations are deemed by Kant not to be relevant in determining our basic moral obligations (even though they may be relevant in applying particular moral laws to concrete cases).[9]

What is so devastating about Kant's view is the way that it enforces the most rigid and exclusionary *fact-value dichotomy* and then proceeds to deny the relevance of empirical knowledge to moral reasoning. Kant insists that the normative character of moral judgments, principles, and values can *only* be derived from pure practical reason. We are thus presented with two entirely distinct realms of norms, some non-moral and the others moral. The former are those that come from our body-based and empirical needs and wants, while the latter issue from an allegedly pure reason that transcends the contingent characteristics of any finite human creature. Armed with this absolute dichotomy, Kant is thus able to sweep away all forms of consequentialist moral theory, because no unconditional morality can be dependent on any contingent human needs or desires. Within this Kantian framework, no need or want that you are given, either by your evolutionary heritage or on the basis of any contingent desires you might have acquired, can be the source of moral laws, although such ends can support hypothetical imperatives: "*If* you want [will] X, then you ought to do Y." This taxonomy of types of imperatives allows Kant to identify "heteronomy," in which the rule governing one's action comes from "the other" (e.g., God, another person, the state, or even one's own bodily desires), as "the Source of All Spurious Principles of Morality" (1785/1983, 441). Not surprisingly, it is empirically based principles (including hypothetical imperatives) that come in for the harshest criticism:

Empirical principles are wholly unsuited to serve as the foundation for moral laws. For the universality with which such laws ought to hold for all rational beings without exception (the unconditioned practical necessity imposed by moral laws upon such beings) is lost if the basis of these laws is taken from the particular constitution of human nature or from the accidental circumstances in which such nature is placed. (442)

With Kant's fateful move, the die is cast, resulting in more than two centuries of mainstream Western moral philosophy predicated on the alleged uniqueness of moral judgment and the radical separation of hypothetical from categorical imperatives. Kant's de-theologized Christian framework is one in which moral principles must have a transcendent source (namely, universal reason), and acting in a morally worthy fashion is conceived as being bound only and absolutely by values and imperatives that you autonomously give yourself, as an expression of your essential rational nature. Supposedly, you can be truly free only when you are bound by moral law you give to yourself; that is, only when you will autonomously.

A "naturalistic" ethics, on Kant's view, is thus an oxymoron. There can be no such thing, other than a merely descriptive study of cultural practices and institutions. Moreover, empirical knowledge cannot be relevant to establishing the normative force or content of our moral principles. We are left with a very strong form of non-naturalism, in which moral judgment is a *sui generis* activity of pure practical reason, independent of any heteronomous determining factors. It is this type of non-naturalism, along with its attendant conception of a unique type of moral judgment, that I will argue is entirely out of touch with our current understanding of how the mind works.

Kant's Unfortunate Legacy

The great and abiding positive value of Kantian moral philosophy is its insistence on respect for the autonomy and well-being of oneself and others. I certainly agree that it would be a far better world if we could all practice the respect toward others that Kant regarded as owed only to rational creatures.[10] Kant's reconstruction of Judeo-Christian morality on rational grounds has much to offer us concerning how we ought to treat ourselves and others. Nonetheless, the great negative counterweight to these important insights is Kant's utter dismissal of hypothetical judgment as a form of moral reasoning, and his denial of any normative role for empirical knowledge.[11] Having taught a course on Kant's theory of morality numerous times over the past thirty-five years, I never cease to be taken aback by his refusal, based on his rejection of any "impure" elements, to recognize that moral values and moral principles can and should arise from our ongoing communal experience of moral disagreement, cooperation, argument, and experimentation. I do not deny that attention to Kant's *Metaphysics of Morals* and his *Lectures on Ethics* reveals a less formalist and purist perspective, and shows Kant's understanding of how difficult actual moral judgment is, once we move away from the purity of the categorical imperative to the messy details of the concrete situations to which we must apply our moral laws. Still, Kant is a purist about the nature and foundations of moral law. In assuming the pure/impure dichotomy, Kant was a man of *his* times, and he apparently still continues to be regarded by many philosophers as a man for *our* times, too, insofar as we are the inheritors of his moral absolutism and purism. In short, he has bequeathed to us a set of highly problematic assumptions, including (1) a classification of types of experience and judgment, (2) a characterization of moral laws as unconditionally binding, (3) an insistence that nothing but pure practical reason can possibly be the source of morality, (4) the strident denial that moral thinking might be a form of transformative problem-solving, and (5) the dismissal of empirical knowledge from the foundations of morals. In what follows, I intend to challenge these foundational assumptions as a way to open up a quite different conception of the nature of human morality as situated, embodied, and imaginative.

Conclusion

The upshot of the previous sections is that Kant's view of moral norms leaves us with a misleading way of rigidly dividing our experiences into independent kinds and then demanding that each kind have its own unique foundational norms. Kant denies the crucial role of empirical knowledge in moral problem-solving, and he tends to reduce hypothetical reasoning to means-ends calculations, as in utilitarianism, where the desired end is supposedly *given*, and then one need only determine how best to achieve that end. On his view, reason is *brought to* experience to impose norms, which, Kant insists, can never be derived from experience.[12]

In sharp contrast, ethical naturalism argues that there is no other source of moral norms than experience, when experience is taken in its richest, deepest sense. Therefore, it is necessary to take a look at how an ethical naturalism understands the sources of our ("moral") values.

TWO

Where Are Our Values Bred?—Sources of Moral Norms

Tell me where is fancy bred,
Or in the heart, or in the head?
SHAKESPEARE, *THE MERCHANT OF VENICE*, III.ii

The central thesis of the previous chapter was that—contrary to the basic assumptions of many mainstream moral theories, and especially those influenced by a Kantian framework—*there is no distinctively and uniquely moral experience, and so there is no need for a unique source of moral norms to account for allegedly unique moral judgments*. The alternative naturalistic view I shall sketch is one that regards experience as multidimensional and not carved up into discrete and independent types. Consequently, there is no need to seek out a special class of distinctively moral norms upon which to ground moral appraisals and judgments. A naturalistic approach conceives of moral deliberation as just a form of problem-solving—an experientially based moral inquiry in which the relevant values grow out of our shared needs, desires, interests, and practices.

Most of the moral deliberation that we engage in arises in situations where we have incompatible values and interests, or where our values conflict with those of others. Therefore, in preparation for my account of moral cognition as a process of problem-solving that works out tensions and conflicts in moral values, I want first to address briefly the issue of where our values come from, if they do not drop down upon us like manna from heaven (that is, from some transcendent source). I need to identify the basic types of human moral values that arise from our nature as biological-cultural organisms. Surveying the range of basic human values that constitute human well-being sets some of the constraining parameters for any adequate human morality, although only in a very loose way that supports a pluralistic view of possible moral systems. Here I make no pretense to exhaustiveness, but only strive to indicate in a general way the basic categories of values that tend to appear in cultures around the world and that we would expect to figure prominently in any moral system. This survey will lead me eventually to conclude that our most significant problem in moral theory is not where our values come from, because, as will be evident, it is easy to catalogue all sorts of values. Rather, the more important issue is how we should decide which values take precedence in situations where there are conflicting and seemingly incompatible values and standards in play.

What Is the Nature of Value?

First, what do we mean by "value"? In ordinary discourse, we often have an unfortunate tendency to speak as if values were objects, commodities, or resources that we can acquire or lose through our actions. For example, we speak of "gaining" or "earning" respect, attention, or love. We can also "lose" someone's respect, and then perhaps try to "win it back." We talk of "seeking" happiness, "finding" happiness, and "throwing away" our happiness—as though it were a valuable commodity for our use. Although this widespread hypostatizing tendency usually creates no serious problems in the ordinary discourse of daily life, it can be quite misleading when we want a deeper, more philosophically adequate sense of what values are and how they affect our lives.

Values are not quasi-objects. Instead of talking about *a value* (as a noun), it is more appropriate to focus instead on the activity of *valuing* (as a verb). Valuing is something we *do*—an activity or dynamic process in which we are directed toward achieving a certain existential state. To speak of something as "valued" is to say that some organism has a predisposition to seek to realize a certain state of affairs, namely, the "valued" or desired state of the organism-environment interaction. Values are therefore *relational*, because they require a relation between an organism and the environments with which it actively engages. Some state of affairs is valuable *for* or *to* some organism, animal, or person. Consequently, properly speaking, *nothing is valuable "in itself"* (i.e., independent of any relations), but only in relation to how it serves a living purposive organism or group of organisms.

In everyday speech we often distinguish between things or states of affairs that are valuable for some purpose (i.e., extrinsic values) and those that are valuable "in themselves" (i.e., intrinsic values). It is relatively unproblematic to speak of something as being "valued for its own sake," if by that locution we mean only that, within a given context, we are directed toward the realization of a certain state of affairs without presently regarding it as a means to some other end. However, it is often erroneously assumed that things valued "for their own sake" are valuable *in themselves* (i.e., unconditionally valuable). This is the sense of the term that Kant was employing when he spoke of certain states of affairs as being "good in themselves, without qualification." *That* conception of absolute values was a significant mistake that proved fateful for much subsequent moral philosophy. It is a serious mistake because nothing is valuable *in itself*, but rather only in relation to a certain type of desiring, striving, active organism. Therefore, I think it best simply to avoid the term "intrinsic value," because it suggests a value given innately, prior to any development of the organism in and through activity within its environment.

Consequently, the only plausible sense of the term "intrinsic value" would be a case that met the following conditions: (1) an organism develops its current biological and cultural makeup as the result of the evolutionary history of its species; (2) given that it also develops ontogenetically through "normal" learning patterns, according to "normal" time frames of sequential development, the organism will come to have certain biological, interpersonal, and social needs that must be satisfied for it to continue to exist and flourish; and (3) under such "normal" developmental conditions, the organism will therefore be predisposed to seek certain organism-environment states.[1] In other words, the organism will have certain values it seeks to realize. Some of these values may be consciously entertained and pursued, but most of them will function automatically within the organism and operate beneath the level of conscious awareness. The problematic terms "innate" and "intrinsic" are probably not the best terms to describe such valuing, insofar as these terms often suggest a biological (genetic) givenness that does not require any experience or activity for its realization. Such a view mistakenly places the value either entirely within the organism, or else it projects it into some realm of Platonic Forms that are supposedly complete in themselves, whereas values actually emerge only as the result of certain relational patterns of organism-environment interaction.

As a way of avoiding terms like "innate" and "intrinsic," I would like to appropriate Robert McCauley's distinction between cognitive capacities that exhibit "maturational naturalness" and those that manifest "practiced naturalness," and I want to apply this distinction to acquisition of values. Instead of talking about innate capacities, McCauley distinguishes "maturationally natural" cognitive capacities, like chewing and walking, from what he calls "practiced" capacities, such as riding a bicycle or writing. Maturational (or maturationally natural) capacities are characterized by the following features:

(1) No one invented them. (2) Their emergence never depends upon artifacts (though some cultures have developed artifacts to facilitate their acquisition). (3) Humans undertake them spontaneously. (4) A few have general forms that we share with other species. (5) Their acquisition does not depend upon explicit instruction or specially structured learning environments, and it does not turn either on inputs that are particular to a culture or (even) on inputs that are culturally distinctive. (2011, 29)

I propose to extend McCauley's notion of the "maturationally natural" capacities to the idea that there are certain maturationally natural *values* shared by nearly all human beings, and some of which are also shared with other animal species. A maturationally natural value—such as the maintenance of dynamic equilibrium of an organism, its ability to sustain the basic conditions of life, or its need for interpersonal engagement—would satisfy all five of the criteria just cited above. Values of this sort will arise naturally for any organism with a certain biological makeup that developed according to a typical plan of maturation.[2]

Furthermore, just as there are "practiced" capacities, there are also values acquired through culturally mediated practices, such as strictures on ethically appropriate forms of dress in certain culturally defined contexts. Both types of values (maturational and practiced) will tend to appear in all members of a species that share a common set of biological capacities (for perception, motion, action, and feeling) and also share patterns of enculturation. Because of the maturational character of these shared values, it would therefore be better *not* to use the term "intrinsic" or "innate" as a modifier for "value" at all, because such usage gives rise to the nearly inescapable tendency to think of values as existing by themselves, independent of the maturational processes of living creatures, or at least prior to any developmental experience of the organism, as if they were built into the organism as essential givens.

As mentioned above, I follow Dewey in preferring the verb "valuing" (instead of the substantive "value"), in order to emphasize the dynamic, directed activity of an organism seeking to function well within its environments. In short, an organism's preferential directedness toward the realization of certain states of organism-environment interaction is an evaluative process, and the "states" toward which the organism is preferentially directed can be selectively and abstractly described as that organism's values, as long as we refrain from turning those "values" into abstract entities. Finally, many of these values will be maturationally natural, while others will be practiced (perhaps culturally specific) values.

This view of valuing as an activity of an organism in its relation to, and engagement with, its environment is perfectly compatible with the claim that some states of affairs are "valued for their own sake." As we observed above, this phrase would mean only that, within a particular setting or context, there will be states of affairs that an organism seeks to realize, which, *at that moment*, are valued by it as consummations of meaning or experience, rather than being taken as a means for achieving some further end. However, a state of affairs that is "valued for its own sake" (i.e., what is colloquially called an "intrinsic" value) is not some timeless, transcendent, non-relative value, because value *always* involves a relation between an organism and its environment(s). Moreover, we sometimes come to see that a value that we originally thought we were pursuing for its own sake, turns out, upon later review, to have been part of a larger course of action, so that it is actually now seen as conducive to some other end(s). It was for this reason that Dewey argued that all ends are actually "ends-in-view," rather than fixed, terminal ends (1922, chap. 19).

The upshot of this examination of the nature of valuing is that we need to disabuse ourselves of the idea that there are absolute values existing independent of any creatures for whom they are significant. Properly speaking, nothing is "good in itself," apart from any relation to a valuing organism, even though there are, within a given context, sometimes ends that we do not currently see as means to some further end.

Basic Sources of Values

With this understanding of the dynamic, relational, interactive nature of valuing, we have the appropriate context for briefly exploring the chief sources of our values. Where do our values come from? The naturalistic answer is that our values emerge within our natural situation (where cultural values are just as "natural" as bodily imperatives arising from our biological needs). Consequently, a genealogical account of values of the sort I am offering here in no way requires any sort of transcendent source. The basic naturalistic argument runs as follows: Human beings are complex, multifunctional organisms in continual interaction with their environments. We are an intricate series of ongoing interactions between our developing biological/social nature and the environments we inhabit, which are at once physical, interpersonal, and cultural. Human creatures seek to survive and grow, and all of this activity naturally occurs in an interpersonal and cultural context. Many of our deepest values come from what is required for the mere survival of the organism, such as the goods of food, water, shelter from the elements, and protection from physical harm. But there are also the more socially constituted goods, arising from our social nature as creatures who exist always in interaction with others, who need the love and care of others, who compete with others for resources, and who thrive via cooperative activity with others. All of these values are values of human flourishing. They do not require transcendent foundations. They are presumptive goods for us, and our problems come when these goods conflict in a given situation, or when our pursuit of some good interferes with another person's pursuit of some good.

There are multiple sources or levels of values that acquire moral weight for the kinds of creatures we are, as we interact with the kinds of environments we inhabit. The person we aspire to be, the goods we pursue, and the situations we seek to realize are all situated within our ongoing organism-environment interactions. Because of the complexity of our bodily organism, the complexity of our interpersonal relations, and the complexity of our environments, there come to be at least the following four levels of values that have an influence on our moral reasoning.[3]

1. Values Tied to Organic Functioning and Well-Being

Our first priority as living creatures is to go on living and to grow. Unless we continue to live and develop, there is not even a possibility of achieving higher levels or forms of well-being. As Owen Flanagan has so nicely phrased it, there is a natural imperative: "fitness, then flourishing" (2007, 54ff.). Flanagan spells out this two-phase dependency relation as follows:

Our innate conative constitution orients us categorically to seek fitness. This innate conative constitution is not best described as consisting of a set of beliefs—it is not epistemically driven. Nor is it, as a given in the first instance, either rational or irrational. Like induction, it is arational. We do not rationally adopt the aim to be fit or to

CHAPTER TWO

use induction. . . . Next comes the magic trick: At every place on Earth, as fitness is achieved (more likely, as it is being achieved), humans begin to strive for meaning and happiness. The quest to flourish invariably reveals that we are possessed of (in virtue of the kind of social animals we evolved to be) a platonic orientation to discover what is good, true, and beautiful. (55–56)

At least some minimal level of organism fitness is an evolutionary imperative for all humans. We have evolved organic bodies that require certain basic conditions for the continuation of life processes, and so we will necessarily come to have certain embodied values that operate for us regardless of whether or not we consciously entertain them. The vast majority of the time we have no awareness of how fitness values are structuring our ongoing experience and action. Indeed, the working of these values automatically and nonconsciously is necessary for our survival and well-being, for if we had to consciously manage all of these bodily events and adjustments, our consciously monitored cognitive systems would overload and lose their ability to manage the conditions of life.

In a series of important books, Antonio Damasio (1994, 1999, 2003, 2010) has explored the biological basis of our deepest and most universal values. Here is Damasio's basic idea:

I see value as indelibly tied to need, and need as tied to life. The valuations we establish in everyday social and cultural activities have a direct or indirect connection to homeostasis. That connection explains why human brain circuitry has been so extravagantly dedicated to the prediction and detection of gains and losses, not to mention the promotion of gains and the fear of losses. It explains, in other words, the human obsession with assignation of value.

Value relates directly or indirectly to survival. In the case of humans in particular, value also relates to the *quality* of that survival in the form of *well-being*. (2010, 47–48)

So, what are our deepest values? They are whatever is necessary for the continuation and growth of our life as organisms. In *The Feeling of What Happens* (1999), Damasio identifies the chief conditions for the maintenance of a living organism as follows: an organism must establish at least a minimal permeable boundary that separates it from, and simultaneously keeps it in contact with, its environment. That boundary then defines what Damasio, following Claude Bernard, calls its "internal milieu." Without a boundary, there would be no organism integrity, but the boundary must also be somewhat permeable, so that food (energy) can be taken in and waste can be expelled. The organism must also continually maintain a relative stability of that "inner" environment, and Damasio borrows Walter B. Cannon's term "homeostasis" to describe the balance of processes necessary for the maintenance of life. Damasio sums up this ultimate biologically grounded value: "For whole organisms, then, the primitive of value is *the physiological state of living tissue within a survivable, homeostatic range*" (2010, 49; italics in original).

Jay Schulkin prefers the term "allostasis" over "homeostasis" (from "allo": change, and "stasis": stable/same) to better capture the dynamic, though relatively stable, character of the processes of life maintenance. Allostasis is "the process by which an organism achieves internal viability through bodily change of state, which comprises both behavioral and physiological processes that maintain internal parameters within the limits essential for life" (2011, 6). Phan Luu and Don Tucker explain that in homeostasis the organism establishes a set point and develops mechanisms for returning to that set point (state of the organism) whenever the original homeostasis is disrupted: "Homeostasis mechanisms respond to deviations from a set-point by initiating restorative responses to correct the deviations and return the system to its homeostatic set-point" (2003, 125). Automatic mechanisms for cooling the overheated body and bringing its temperature back within an acceptable range would be a good example of a homeostatic system.

Allostasis, in contrast, makes it possible to actually change the set-point values in response to changes in the organism and environment.

Allostasis can be defined as the regulation of many variables (including behavioral as well as physiological variables) over time to meet motive set-points that are established dynamically. . . . Rather than just responding to deviations from fixed set-points, allostasis allows the organism to anticipate needs and to adjust configurations of regulatory goals in advance to meet those needs. . . . In essence, allostasis incorporates learning and adaption [sic] into the process of homeostatic self-regulation. (Luu and Tucker 2003, 125–26)

It appears to me that when Damasio uses the term "homeostasis," he is often (though not always) including the more dynamic processes of allostasis described above, so, henceforth, I will often use the term "allostasis," rather than "homeostasis."

This ur-value of dynamic biological allostasis generates a number of subsidiary biological values that are essential for any human notion of well-being. A non-exhaustive list of these universal biological values would include at least the following:

CHAPTER TWO

- food capable of sustaining life processes
- a moderate climate that avoids the extremes of heat and cold that would destroy life
- sufficiently clean air and water
- protection from predators or others who might do one harm
- care and nurturance necessary for an infant (and child) to traverse at least most of their developmental trajectory to the point where their physical, cognitive, and emotional capacities are well formed
- bodily systems (e.g., the immune system) to deal with illness, bodily injury, or psychological dysfunction

One might extend this list of basic biological values, but the present list is sufficient at least to make clear the broad range of enabling conditions that must be in place for even the most modest human survival and well-being. It is not surprising, therefore, that in what is known as the "Capabilities" approach to global justice (Nussbaum 2000), these and other life-maintenance values tend to show up on lists of basic conditions that, it is argued, we are morally obligated to establish, because people everywhere need these in order to survive and flourish.

Damasio nicely sums up his basic contention about the ultimate biological origins of our values:

My hypothesis is that objects and processes we confront in our daily lives acquire their assigned value by reference to this primitive of naturally selected organism value. The values that humans attribute to objects and activities would bear some relation, no matter how indirect or remote, to the two following conditions: first, the general maintenance of living tissue within the homeostatic range suitable to its current context; second, the particular regulation required for the process to operate within the sector of the homeostatic range associated with well-being relative to the current context. (2010, 49)

2. Values Tied to Intimate Interpersonal Relations

Our survival and well-being are not just matters of physical sustenance and protection. They are equally matters of social (interpersonal) interaction too. We are not born autonomous agents who then choose whether, and to what extent, we will enter into social arrangements or interact with others of our species.[4] Instead, our sociality is an evolutionary necessity, an unavoidable condition of our very existence. We are fundamentally social creatures whose physical, psychological, emotional, and social well-being rely mostly on the goodwill and care of others, especially for the first two decades of our lives when we are developing our biological and social capacities, and also often at the end of life (Stern 1985; Sheets-Johnstone 2008). Because of this, values and norms emerge that are connected to our need for nurturance, care, education, and social interaction. This radical developmental dependency includes every aspect of our being, from our physical capacities for perception and action, to our emotional functioning, to our ability to engage in full-blown socially cooperative behaviors.

Our social nature is not a theoretical perspective, nor is it a reflective stance we take toward others. It is, rather, what Colwyn Trevarthen (1979) calls our "primary intersubjectivity." An infant is born primed to engage immediately in an amazingly choreographed communicative dance with its caregivers. This primary give-and-take is a form of shared intentionality, in which the interacting parties share with each other control of their intentions, that is, their directedness toward objects and goals. Trevarthen's many studies during decades of research reveal how infants can imitate the facial and bodily gestures of their primary caregivers, engage in turn taking in their communicative interactions, mirror the other's emotional states, communicate their basic needs and intentions, and so on. All of these capacities and activities are geared toward the bonding and cooperative behavior that are necessary for the survival and well-being of the infant or child.

One essential condition for social engagement is the understanding of other people as minded individuals who experience consciousness, thoughts, and feelings, just as you do. The field known as "theory of mind" research focuses on how it is that infants come to have a sense of the other as possessed of "mind" and various "mental states and attitudes" that involve limited and perspectival knowledge. There are two dominant perspectives on these phenomena, which have been christened the *theory* theory and the *simulation* theory. Roughly, the *theory* theory argues that children learn about other minds by employing something like a theoretical stance in order to interpret the perspectives and behaviors of others (Gopnik and Meltzoff 1997). The *simulation* theory regards the *theory* theory explanation as far too intellectualized and reflective, and argues instead that children assume that others' minds are analogous to their own and therefore they simulate the thoughts and feelings of others, as if they were experiencing the same situation as the other (Gordon 1986; Goldman 1989).

Although both of these popular theories have been valuable in directing our attention to our basic social interconnectedness, Shaun Gallagher and Daniel Hutto (2008) observe that neither of these views captures the pre-reflective depth of interactive engagement that is the mark of primary intersubjectivity. Most of the time, we are not egos trying to "figure out" what is going on in the minds of others, as if we observed their behaviors and then theoretically explained those to ourselves. Nor are we merely simulating a behavior in ourselves and assuming, by analogical projection, that another person's behavior must be like ours. *Rather, we are simply intersubjectively co-active with others in creating a shared communicative and performative situation.* It might be said that, at this level, we exist immediately in and through others, via a nuanced interplay of our intentions and points of view. Yes, there is simulation, and, yes, there is even primitive theorizing on occasion, but mostly there is our dwelling with and through others in a mutual interplay of shared intentionality and coordinated behavior.

Because we share in primary intersubjectivity, we acquire certain values that make this form of mutually coordinated engagement possible. Among the more important values of interpersonal relations and sociality are the following:

Care/Nurturance

To state the obvious, without a considerable degree of care and nurturance by others over a very extended period of time, an infant or child will fail to develop the "normal" functional capacities necessary for life and growth. The care we need is so comprehensive, so labor intensive, and so psychologically draining on the caregiver, that the fact of its existence is somewhat a minor marvel of the human species. The psychological cost of the nurturance basic for life maintenance is so great that some evolutionary psychologists have felt it necessary to devote their careers to explaining how care of offspring, and even care of others not related by blood, is possible within a broadly self-interested perspective. We need not enter into the "selfish gene" debate to validate the fact that care and nurturance of others is a fundamental value of all cultures, even though the requirements of appropriate types and levels of care can vary widely across cultures.[5]

Why do self-interested creatures like us typically care for others, even at the expense of our own comfort, satisfaction, and well-being? Patricia Churchland has examined the evolutionary and developmental dimensions that tend to make us care not just for ourselves and our individual well-being, but also for the well-being of others. She begins by observing that "*caring* is a ground-floor function of nervous systems. Brains are organized to seek well-being, and to seek relief from ill-being" (2011,

30). Humans (and some other animals) have evolved to extend caring from attention to our own welfare to that of selected other conspecifics. Churchland summarizes research revealing the crucial role of neuropeptides like oxytocin and arginine vasopressin in pair bonding and nurturing behaviors within certain mammals, such as prairie voles: "A compelling line of evidence from neuroendocrinology, which studies hormone-brain interactions, suggests that in mammals (and quite possibly social birds), the neuronal organization whereby individuals see to their own well-being was modified to motivate new values—the wellbeing of certain others" (14). She sees attachment, by which she means "the dispositions to extend care to others, to want to be with them, and to be distressed by separation" (16), as crucial for all "normal" human development and as providing the basis for morality. In short, Churchland argues that "morality originates in the neurobiology of attachment and bonding, depends on the idea that the oxytocin-vasopressin network in mammals can be modified to allow care to be extended to others beyond one's litter of juveniles, and that, given that network as a backdrop, learning and problem-solving are recruited to managing one's social life" (71).

In *Why Good Is Good* (2002), Robert Hinde also observes that humans tend to manifest *both* self-interested and other-directed dispositions, emotions, and behaviors. He begins his list of universal moral principles allegedly found in all cultures and throughout all of human history with the imperative that "parents should look after their children" (12). However much this seems like an excellent candidate for a universal principle grounded in our basic human need for care and empathy, the obvious problem is what "look after" means within different cultures. Presumably, it would include at least the following: (1) supplying life-sustaining nutrition, (2) protecting the child from harsh natural elements and from bodily harm that would damage their capacities for life maintenance and growth, (3) expressing at least a minimal degree of affection and love, and (4) establishing some form of educational process that helps the child develop the life skills necessary for them to survive and then to pursue goals constitutive of some sense of well-being and flourishing (or even "happiness"). Expectations about the obligation to foster development in one's offspring beyond such minimal requirements are probably culturally relative. My point here is that some level of human affection, care, and nurturance is a necessary requirement for virtually any conception of *eudaemonia* (well-being, flourishing) that we can imagine.

What has come to be known as the "ethics of care"—often traced to Carol Gilligan's *In a Different Voice: Psychological Theory and Women's Development* (1982)—argues that nurturant care and responsibility for others are values equal in importance to the values of rights, justice, and duty that Lawrence Kohlberg (1981, 1984) discovered in his groundbreaking studies of children's moral development. Based on his studies of boys at the primary school level, Kohlberg had postulated six major stages of moral growth, each of which manifested a different perspective on the perceived sources of moral obligation. Gilligan, who had worked with Kohlberg on some of these earlier studies, expanded the subject pool to include girls, and she then noticed their typically greater emphasis on care for others and maintaining social harmony and cooperation. Gilligan was careful not to draw the *rights and justice* versus *care and nurturance* orientations strictly along gender lines, observing only that girls tended to be socialized around conceptions of their role as caregivers more than boys did. Nell Noddings (1984) argues for mothering as a model for the nurturant care that ought to provide the foundation of any moral system.

Some versions of care ethics have subsequently been criticized for their tendency to overemphasize self-sacrifice at the expense of self-interest, thereby reinforcing oppression of caregivers (Hampton 1993), and for certain psychologically and sociologically unrealistic assumptions about mothering and the nature of nurturance (e.g., LaChance Adams 2011),[6] but care ethics has now become a mainstream orientation that complements rights/justice/duty approaches and claims to provide a model for all of our moral interactions (Held 2005; Kittay 1999).

The caring that is necessary for life maintenance and growth cannot be limited only to basic life-sustaining conditions, however primary that task must always be. It must extend to the development of human intellectual, emotional, aesthetic, and social capacities that (1) make it possible for a child to use its own developing capacities to direct its life activities, and (2) to enter in an appropriate manner into social relationships ranging from the most intimate bonding to membership in larger, but less personal, social groups.

Empathy

The care that is necessary for survival and flourishing is for the most part predicated on the possibility of empathy, which is our capacity to experience the situation of another person—an ability to feel with and for them. Franz de Waal defines empathy as "the capacity to (a) be affected by and share the emotional state of another, (b) assess the reasons for the other's state, and (c) identify with the other, adopting his or her perspective. This definition extends beyond what exists in many animals, but the term 'empathy' . . . applies even if only criterion (a) is met" (2008, 281). In *A Treatise of Human Nature* (1739), David Hume famously declared that morality was founded on feeling (i.e., passions and sentiments), rather than reason, and he identified what he called "sympathy" (as a feeling for the well-being of others) as the foremost moral emotion, upon which the very possibility of morality is founded:

> Most people will readily allow, that the useful qualities of the mind are virtuous, because of their utility. . . . Now this being once admitted, the force of sympathy must necessarily be acknowledg'd. Virtue is consider'd as means to an end. Means to an end are only valued so far as the end is valued. But the happiness of strangers affects us by sympathy alone. To that principle, therefore, we are to ascribe the sentiment of approbation, which arises from the survey of all those virtues, that are useful to society, or to the person possess'd of them. (1739/1888, 618–19)

Hume's sympathetic fellow feeling requires our empathic ability to feel for and with the other in relation to their state of satisfaction or dissatisfaction, but it goes beyond empathy to an active concern for the well-being of others. Hume was rightly recognizing that the core of any system of moral relations involves, first, a basic empathic understanding of the pain and pleasure, suffering and thriving, failing and flourishing of other human beings, and, second, the necessity of caring about these things in the lives of others. Hume's focus on feeling relations toward others is thus in direct contradiction to the Kantian claim that only an allegedly pure practical reason can be the appropriate source of moral action.

Not surprisingly, empathy has today become one of the central foci for current naturalistic approaches to ethics. Starting in the early 1980s, Martin Hoffman wrote a series of articles on the central role of empathy in moral development, and he did this at a time when mainstream Western moral theory was still mostly under the sway of the Kantian claim that feelings cannot provide an adequate basis for morality (Hoffman 1987, 2000). Today empathy is routinely touted as *the* primary emotional relation that underlies our moral interactions, and it is often taken to be the key to the possibility of an ethics of care (Gilligan 1982; Noddings 1984).

The discovery of the so-called "mirror neuron" system in humans and certain other species has breathed additional life into the empathy movement. This research began when it was discovered that a macaque monkey's seeing another monkey or human grasp an object (say, a peanut or a banana) activated parts of their motor cortex that would be activated if the monkey were to have performed that same grasping motion (Gallese et al. 1996). In other words, there appeared to be a sensory-motor simulation process involved in certain primates seeing certain very specific actions performed on or toward specific objects. Further research expanded the types of simulation involved (e.g., not just vision, but also with hearing and touching) and also recognized fairly precise constraints on which actions generated the mirror-neuron system's firing (Rizzolatti and Craighero 2004). Soon this research extended to humans, and it now appears that a mirror-neuron system is operative not just in seeing a specific action performed, but also in hearing or reading the word for that action, imagining the performance of that action, and dreaming about such a performance (Decety and Grezes 2006; Aziz-Zadeh et al. 2006). It has now been hypothesized that the mirror-neuron system is responsible for our capacity to respond empathically toward others (Decety and Jackson 2004; Gallese 2001). For most humans, when they see or hear you in severe pain, or see or hear you crying, they "feel with" you in an intimate way.[7] This empathic experience is not a theoretical hypothesis about another's behavior, nor merely an analogical projection from my experience to yours; rather, it is an activation of sensory, motor, and affective processes in direct relation to the behavior I perceive in you.

However promising the mirror-neuron hypothesis might seem as a way of accounting for our grasp of others' intentions, for our imitative behavior, and for our capacity for empathy, Patricia Churchland (2011) has given us ample reason to be cautious, if not downright skeptical, about many of the boldest claims that mirror neurons are the key to all moral relations. She surveys a large number of studies and then, for each type of study, outlines the chief problems with drawing any clear or strong conclusions about the workings of a mirror-neuron system in humans. There is some promising evidence, but for the present we appear to be far away from any conclusive account. What seems defensible at this point is basically that the vast majority of humans are capable of empathy and other associated behaviors, such as social mimicking, and this appears to depend on some role for our mirror-neuron capacities.

3. Values Tied to Complex Social Interactions and Institutions

What begins with interpersonal engagement (one-on-one, face-to-face encounters) eventually extends to activity in larger social groups, giving rise to what are known as the social virtues. Social groups and institutions are necessary for orchestrating our communal relations so that we can harmoniously pursue both our individual and our communally shared ends. As Michael Sandel (1998) and many other communitarian political theorists have argued, contrary to overly simplistic *social contract* theories of society, humans are not born independent individuals who then freely choose all of the social arrangements they will enter into. Instead, we are social creatures from the very beginning of our life outside the womb (and probably within it)—from our primary intersubjectivity up through our participation in large-scale social groups. The point here is not that we do not sometimes choose the groups of which we will be a part (e.g., clubs, neighborhoods, religious organizations, schools, political parties, social organizations, social movements, and so forth). Rather, the point is that we are inherently social creatures, such that our moral identity cannot be extricated from our mostly inescapable webs of social interaction.

Our social nature thus requires of us certain virtues—dispositions of character—that are necessary for social cohesion, harmony, and cooperation. As Alasdair MacIntyre has argued, "A virtue is an acquired human quality the possession and exercise of which tends to enable us to achieve those goods which are internal to practices and the lack of which effectively prevents us from achieving any such goods" (1984, 101). Recall, from chapter 1, MacIntyre's notion of a practice as a complex form of socially established cooperative activity that gives rise to goods that are achievable within that practice and that generates notions of excellent performance involved in trying to realize those goods. Specific virtues are thus valuable just insofar as they make excellent practices possible. The relevant practices are given form by the history and traditions of particular group endeavors, and the social virtues are specified by the nature of those endeavors.

If there are any virtues necessary for all types of practices, they would be quite general in nature. However, although MacIntyre does recognize a small number of candidates for universal virtues (such as courage and truthfulness), he argues that there will always be multiple moral traditions of the virtues, each tied to historically specific places and times. Therefore, in many cases, these historically situated traditions will give rise to different internal goods and generate their own catalogue of primary virtues directed toward those goods of practice. Quite often there will be virtues central to one tradition that do not even make the list of virtues in another cultural system. For example, humility, deemed

CHAPTER TWO

a supreme virtue within Christianity, would have little or no place in Homeric society with its primary emphasis on courage, cunning, and loyalty. Achilles may be excessively proud, but none of his compatriots expects their leader in battle to be truly humble, whereas within Christianity, humility is a fundamental condition for even the possibility of moral goodness. Nor would we expect there to be a significant counterpart in Homeric society to the industriousness so valued in Ben Franklin's America. In short, a comprehensive list of virtues would be quite long, one would not expect to find more than a handful of virtues valued in all societies, and even then those universal virtues would most likely be understood somewhat differently in each cultural setting. Nevertheless, perhaps we can propose some representative social virtues that tend to come with virtually any conception of social organization.

- *Truthfulness.* Immanuel Kant spilled buckets of ink trying to articulate and defend his almost fanatical insistence on truth-telling and promise-keeping. His moral epistemology saddled him with the necessity of finding an absolute prohibition against lying that could be derivable *a priori* from the very notion of rational agency itself. However, we do not need Kant's rationalistic purism to realize that no interpersonal relationship and no social organization can survive without some commitment to veracity and promise keeping. Simply put, if you cannot trust and act on what I say and do when I purport to express my views and intentions, then the bonds of social relation and coordinated communal activity are severed. In his *Lectures on Ethics*, even Kant did not need his infamous framework of pure practical reason to justify his reasonable claim that "without truth social intercourse and conversation become valueless. We can only know what a man thinks if he tells us this thoughts, and when he undertakes to express them he must really do so, or else there can be no society of men" (1930/1963, 224).
- *Integrity.* Integrity is a wholeness, unity, and stability of character that makes it possible for others to have a reliable sense of who you are and how you will behave. It can extend to having a firm adherence to certain values or principles of action that are manifest in your behavior. When a person's integrity shows itself in truthfulness and promise keeping, we know we can count on what they say and do, as a basis for how we should act in relation to them. The benefits of this are obvious for any social group. A "weak" character lacks integrity and therefore lacks the ability to maintain a certain degree of unity of character in the face of life's difficulties and challenges.
- *Courage.* Courage is one of the top candidates for a universal virtue, since a reasonable strength to persist in one's actions in the face of threats of death or bodily or psychological harm is virtually a necessity for the preservation of a society's freedom

from oppression and from continuous threats of harm. Life so routinely confronts us with potential harm that some measure of courage is required even for our passage through a single routine day. Given the virtual omnipresence of physical danger and moral evil, every society needs at least some of its members to be willing to take courageous stands on matters threatening the well-being of the community and its members. Beyond physical courage, we need intellectual and moral courage to do what we, or our society, regards as good and right.

- *Acknowledgment of Authority.* In spite of our Western predilection toward unfettered freedom and autonomy of choice, no society can long endure without a least some measure of recognition of binding authority. This sometimes takes the form of obedience to an authority of state (king, premier, president, representative), but today it often takes the form of acknowledgment of the binding nature of laws created to constrain and guide patterns of exchange, forms of social interaction, and large-scale communal transactions. These laws might be thought to come from a divine authority, from the authority of universal reason, or from state authority. Jonathan Haidt (2001, 2012) has even argued that hierarchy and authority constitute one of the five key dimensions of any fully developed moral system.
- *Loyalty.* Josiah Royce defined loyalty as "the willing and practical and thoroughgoing devotion of a person to a cause" (1908, 9), where the cause is not merely one's personal goals, but a cause shared by the public as beneficial to their joint well-being. Royce interpreted loyalty to another person or group of persons to be loyalty to some shared cause. He regarded loyalty to a cause as a necessary condition of our moral transactions and achievements, so much so that he elevated what he called "loyalty to loyalty" to the status of the supreme principle of morality. He claimed that moral actions are those that foster loyalty. We need not espouse Royce's all-embracing conception of loyalty to appreciate the importance of loyalty for the increase of human well-being—from the level of intimate personal relationships all the way up to the most comprehensive communal practices. Loyalty thus links up with some of Haidt's other four determinants of moral systems, such as reciprocity/justice and authority/hierarchy.
- *Civic-mindedness.* Although a sufficiently large society might survive and even flourish with a few of its members committed solely to their own perceived self-interest, a general concern for the well-being of one's basic social institutions and a commitment to socially cooperative actions are prerequisites for the maintenance of a functioning community.[8] This civic spirit is an extrapolation to the larger social scale of the care and concern for the other that is necessary at the interpersonal level.
- *Other Social Virtues.* It is unnecessary to attempt a comprehensive listing of values and virtues emerging at the social level. One could quite easily expand the list to include such apparently non-universal values as industriousness, efficiency, honor, shame, wit, cunning, perseverance, altruism, benevolence, sensitivity, and so forth.

4. Values Tied to Our Quest for Meaning, Growth, and Self-Cultivation

Morality concerns well-being—of ourselves, other people, and, some would argue, even the more-than-human world. For humans who have achieved at least some minimal level of fitness for survival and growth, our well-being will also include our desire for a meaningful, fulfilled existence. This is what Flanagan means when he says that fitness comes first, then flourishing. We have a strong urge for personal growth and meaningfulness in our lives, and this involves a desire for certain types of experiences and satisfactions. As Socrates so eloquently put it, speaking to Crito from his prison cell shortly before his execution, "The most important thing is not life, but the good life . . . and . . . the good life, the beautiful life, and the just life are the same" (*Crito* 48b). Flanagan has suggested that if we could naturalize (de-transcendentalize) Plato's theory of forms, we could then see that his notions of "'the good,' 'the true,' and 'the beautiful' are ways of gesturing at, or describing, the three fundamental and universal ways humans orient themselves in and toward the world in order to live well and meaningfully" (2007, 40). Flanagan's point is that we tend to find meaning for our lives, and we tend to feel fulfilled, when we are realizing our potential as intelligent (knowledge-seeking), moral, and aesthetically directed creatures.

This sense of meaningfulness and well-being (*eudaemonia*) can obviously vary dramatically by degrees, across cultures, among people within a culture, and even over the span of one's lifetime. We do not, and cannot, hum along at a high peak of intense activity, intense meaningfulness, or intense well-being for long periods of time. Life, as Dewey often observed, is an affair of ebb and flow, peaks and valleys, consummations and dissolutions. However, amidst life's ups and downs, we typically retain a somewhat stable need for some kind of meaningful activity to which we can commit ourselves, both day to day and also long term. These activities can range from gardening to preparing a good meal to quilt making to computer games to child rearing to dancing to philosophical reflection on the meaning of life.

Paul Thagard has drawn on empirical studies of happiness to mark the difference between a *meaningful* life—a life that gives us a reason to get up and get going every morning, because we have things we care about, goals to pursue, projects afoot—and a *happy* life, which requires that at least some of these basic life projects allow us to achieve at least some measure of satisfaction or fulfillment:

A meaningful life isn't just one where happiness is achieved through accomplishment of goals, but one where there are worthwhile goals to pursue. Many of the goals that people value most—for example, raising children and working at challenging tasks—are not always sources of happiness. . . . Hence a meaningful life is not just one in which all your goals are satisfied, but one that provides reasons for doing things. Because meaning requires pursuing goals that are not yet satisfied, it cannot be identified simply with the satisfaction of goals as measured by happiness or well-being. A meaningful life is one where you still have something to do, even if doing it may not make you happy that day, week, or year. (2010, 148–49)

Thagard summarizes the contrast between happiness (as a sense of accomplishment and satisfaction) and a sense of meaningfulness (as a dynamic pursuit of goals) when he says that "happiness is not the meaning of life. Happiness is having your goals satisfied, but meaning additionally involves having worthwhile goals that may or may not be satisfied" (149).

What makes people happy, what different cultures mean by "happiness," and what gives people a sense of meaning in their lives are things that can be investigated empirically. There is now a rapidly growing cottage industry on the cognitive science of *eudaemonia* (well-being, flourishing). The dramatic development of what is known as positive psychology has spawned an entire field dedicated to studying the activities and conditions that appear to give people a sense of general well-being and a sense of meaning (Seligman and Chikszentmihalyi 2000; Gable and Haidt 2005).

While not pretending to offer an exhaustive review of the positive psychology literature, Thagard has suggested a summary slogan of the main findings: "The meaning of life is love, work, and play" (2010, 165). We find meaning, though not always happiness, in loving relationships, dedication to work that we care about, and all of the playful activities that happen to capture our fancy. Without these, or at least the prospect of these in some measure, we lose our reasons for going on, we fall ill, we become depressed, we drift, and we sometimes resort to destructive and ultimately unsatisfying behaviors in search of moments of high arousal and temporary satisfaction.

Certain capacities and virtues are conducive to our pursuit of meaning in our lives. Any character trait or disposition that helps us live a more meaningful and fulfilled life comes to be valued for its functional role in the quality of our life. This list of *eudaemonistic virtues* would include many of the values we have already listed at other levels, since we often cannot draw a sharp boundary between supposedly *personal* virtues and

CHAPTER TWO

social virtues. For example, part of living a fulfilled life involves some commitment to developing one's powers of thought, one's aesthetic sensibilities, and one's moral dispositions, and these will also serve us well in our pursuit of biological well-being and equally in our need for social interactions and harmony. Here is a short list of some of the more important virtues that serve our quest for meaning and fulfillment in life.

- *Intelligence.* Our rational capacities are what make it possible for us to critically and creatively inquire into even highly complex life problems that confront us. We need not essentialize or canonize our rationality into a transcendent Reason (capital *R*) to appreciate that it is, indeed, our rationality that has given us a slight evolutionary advantage in understanding our world and devising intelligent ways of engaging and transforming it.[9]
- *Aesthetic Sensitivity.* The "aesthetic" should not be narrowly restricted only to qualities of artworks or what is known as "aesthetic experience." I mean, rather, that the "aesthetic" encompasses everything that goes into the possibility of a meaningful experience, and this includes images, qualities, schemas, models, concepts, feelings, emotions, and larger interpretive structures like narratives (Johnson 2007). Consequently, *aesthetic sensitivity* makes it possible for us to be aware of, present to, and involved in what is happening and what is possible by way of experience and action. An appropriate sensitivity is what enables you to engage others in a morally perceptive manner. It allows you not only to read (i.e., to experience, interpret) the motivations and intentions of others, but also contributes to greater depth and richness of meaning, both in your current situation, and over the course of your life as a whole.
- *Open-mindedness/Flexibility of Thought.* Intelligence can be worthless, or even immoral, if it operates in the service of a closed mind. Fundamentalism—intellectual, moral, or religious—is the single greatest threat to intelligent inquiry into the perennial problems of human existence. I will give the argument for this claim in a later chapter, but for now I simply state that the refusal to submit one's own assumptions and values to the same critical scrutiny that we so readily and enthusiastically direct toward others is one of the great sins that plagues the human race. As a counterweight to cognitive rigor mortis, it is our plasticity of mind, with its corresponding flexibility of thought, that occasionally permits us to rise above the sedimented habits of thought, feeling, and valuing into which we have too comfortably settled. Without this flexibility, we would have no capacity for criticizing and remaking our situation, so we would lack any resources for forward motion when our entrenched habits run up against changing conditions of experience.
- *Creativity.* Open-mindedness is a necessary, but not a sufficient, condition for the degree of creativity that appears to be distinctive of humans. Once again, I am not denying the possibility of creative problem-solving (such as rudimentary tool use) in

other species. However, I trust that nobody would deny that the human capacity for language use (and other forms of symbolic interaction) along with the plasticity of our neural architecture give us an exceptional advantage in our capacity for critical inquiry and subsequent creative problem-solving. The neural plasticity of the human brain, at its present stage of evolution, far outshines that of any other animals, and this is due primarily to the amazing complexity of the neural architecture of our brains, with its layers of neuronal clusters, massively interconnected via one-way and re-entrant loops of neural connections, all operating in parallel over relatively short time windows (Edelman 1992; Tucker 2007; Feldman 2006). The emergence of novelty from these vast interrelated functional systems is our only possibility for coming up with new ideas and strategies when our entrenched patterns of thought and feeling are not adequate to the demands of our changing experience.

Just as with the other three major sources of our values, this fourth source (i.e., values tied to our pursuit of meaning and growth) could be extended to include a broad range of capacities and dispositions that help us to flourish. I am thinking here of perseverance (determination), industry (hard work), sociability, playfulness, and other minor virtues.

Haidt's Six Foundations of Moral Systems

What, at this point, are we to make of this fourfold taxonomy of moral values and virtues? First, it is quite obviously a partial and selective list intended only to capture what I consider to be the chief sources of our moral values. Second, all of these values arise naturally for creatures like us, that is, they are for the most part maturationally natural, and they therefore need no supernatural or transcendent source or grounding. Third, there will no doubt be other equally illuminating ways to arrange a representative catalogue of values and virtues. I see no purpose in attempting to provide an exhaustive list of values, because there are too many different ways of classifying them, and there are too many culturally different interpretations, even of those values that appear to be universal. Edmund Pincoffs (1986), for instance, lists upward of seventy different virtues, and MacIntyre (1984) surveys a large number of virtues that arise within a number of different historical and cultural contexts, from heroic societies to the Aristotelian polis to medieval communities to colonial America, and so on. Were we to include various non-Western traditions, this list might grow substantially (as in Flanagan 2007).

In assembling my list of sources of values, I began with the necessities of our organic bodily maintenance, moved on to the interpersonal (social) constitution of our identity, expanded the interpersonal to the highest and most comprehensive levels of group or communal life, and ended with our need for personal and communal meaning and a sense of flourishing.

There are obviously other ways of drawing the boundaries of the basic values and virtues of the human species. One of the most influential current taxonomies is Jonathan Haidt's identification of six foundations of moral systems around the world and throughout human history. Haidt and Craig Joseph (2004) had earlier listed five of what they take to be the basic adaptive challenges of our social development and life (as proposed in anthropological literature and evolutionary psychology), and they then explore how moral systems apparently arise to meet each of these challenges. They do not claim that every morality will include all five considerations, but they make a compelling case for their five as the primary determinants of moral frameworks that address basic human needs and social arrangements. Here is Haidt's concise summary of the five foundations of moral systems:

- The *care/harm foundation* evolved in response to the adaptive challenge of caring for vulnerable children. It makes us sensitive to signs of suffering and need; it makes us despise cruelty and want to care for those who are suffering.
- The *fairness/cheating foundation* evolved in response to the adaptive challenges of reaping the rewards of cooperation without getting exploited. It makes us sensitive to indications that another person is likely to be a good (or bad) partner for collaboration and reciprocal altruism. It makes us want to shun or punish cheaters.
- The *loyalty/betrayal foundation* evolved in response to the adaptive challenge of forming and maintaining coalitions. It makes us sensitive to signs that another person is (or is not) a team player. It makes us trust and reward such people, and it makes us want to hurt, ostracize, or even kill those who betray us or our group.
- The *authority/subversion foundation* evolved in response to the adaptive challenge of forging relationships that will benefit us within social hierarchies. It makes us sensitive to signs of rank or status, and to signs that other people are (or are not) behaving properly, given their position.
- The *sanctity/degradation foundation* evolved initially in response to the adaptive challenge of the omnivore's dilemma, and then to the broader challenge of living in a world of pathogens and parasites. It includes the behavioral immune system, which can make us wary of a diverse array of symbolic objects and threats. It makes it possible for people to invest objects with irrational and extreme values—both positive and negative—which are important for binding groups together. (Haidt 2012, 153–54)

Haidt later realized the need to identify a sixth major foundation, which he called the *liberty/oppression foundation*, "which makes people notice and resent any sign of attempted domination. It triggers an urge to band together to resist or overthrow bullies and tyrants" (2012, 185). Haidt's six foundations cover many of the same value systems as my four levels of values, but they give more weight than I have to authority and purity/sanctity. I do not see any fundamental disagreement here. In my account of both development within the family setting and in the structure of broader social institutions, authority relations are bound to be extremely important. George Lakoff and I (1999) noted the primacy of authority and obedience within the "Strict Father Family" model of morality and observed that even within a "Nurturant Parent" morality there will be authority relations and a place for obedience and adherence to duties and responsibilities. I regard the sanctity/degradation foundation as emerging at the level of the conditions for the maintenance of ourselves as biological organisms (where disgust reactions and the operation of immune responses is crucial for our survival and well-being) and then being extended in some moralities to notions of purity of will and moral or spiritual cleanliness. In Johnson (1993), I analyzed the metaphor of moral purity, which shows up repeatedly both in Western religious and non-religious moral systems.

In general, I therefore find the six foundations model extremely useful as an explanation for recurring considerations or concerns that appear to arise in nearly all moral systems. It goes a long way toward explaining the evolutionary emergence of moralities that exhibit some common characteristics. However, as I will explain in chapter 4, I take issue with the characterization of these as innate evolved cognitive modules. My criticisms, however, are directed only to the strongest versions of modularity that speak of moral instincts and moral faculties, whereas Haidt (2012, 123–27) is very careful not to overstate the innateness hypothesis. He prefers the metaphor of each moral foundation as comparable to "the first draft of a book that gets revised as individuals grow up within diverse cultures" (153). As I will argue in a later chapter, the postulation of moral modules tends to be innocuous if it is taken to mean simply that humans have evolved a small number of systems for dealing with the kinds of problems and pressures they recurrently tend to encounter in their lives. However, when the alleged modules are defined at this very high level of generality, the notion of modularity seems not to be doing significant work. Moreover, an adequate account of the complexity and variation of systems of morality cannot be explained merely by reference to moral modules (see Flanagan and Williams 2010). There are other claims in Haidt's view with which I take issue,[10] but they are not relevant to my present project, nor do they affect my acknowledgment that Haidt has indeed found at least six very important dimensions of most, if not all, human moralities.

Values as Naturally Arising

I want to draw two conclusions from the previous survey of four basic levels at which we realize our most important values. The first is that all four of these tiers of values can be conceived as having natural sources. For human animals like ourselves, all of these values will emerge to play at least some role in our morality, and they will emerge from the nature of our bodies, the nature of our interpersonal face-to-face interactions, the nature of our participation in larger groups, and our quest for meaning. Because these values can be explained within a naturalistic framework, there is absolutely no need to posit transcendent, pure, or supernatural grounds for any of them.

The second major conclusion is that, because values are a dime a dozen, the source of our values is really not the most important question of morality. Once we see how our values arise naturally from our experience (including our evolutionary history), we will outgrow the need to search for absolute foundations. Our central question will then become how we ought to decide, in particular situations, which values should take priority, especially in the face of potentially multiple values in any given situation that either (a) might conflict with other values we hold or (b) might conflict with values held by other people. That, I submit, is the real moral problem for most of us. It is an issue about the nature of good moral deliberation, and so it is to that question that I now turn.

THREE

Intuitive Processes of Moral Cognition

The Need for a New View of Moral Reasoning

One of the most earth-shattering discoveries to come out of the cognitive sciences over the past three decades is that human thinking and willing operate mostly beneath the level of our conscious awareness, often involving intuitive and highly affect-laden processes. This claim is unsettling because it undermines what many philosophers have previously regarded as the *terra firma* upon which they had erected their entire edifice of theorizing about mind and thought—namely, their conviction that reasoning and willing must be radically free (autonomous), conscious, and deliberate processes of analysis, judgment, and choice. Autonomous reason and will are thought to be what separates human cognition from the causally determined cognition of "lower" animals, affording humans alone in the universe the status of being moral creatures.[1] According to this view, if there is no radically free and self-legislating reason and will, then the very idea of moral responsibility for freely chosen acts evaporates, and along with it goes our traditional conception of morality. This discovery shatters the illusion of a firm ground of transcendent moral reason. It also leads many people to fear, quite mistakenly, that such a view would reduce humans to nothing more than what C. S. Lewis called "trousered apes."

Moreover, moral absolutism typically assumes the traditional notion that humans possess an autonomous reason that discovers, and an autonomous will that chooses whether or not to obey, pre-given universal moral laws. Consequently, if our moral thinking turns out to operate primarily through processes that are intuitive and mostly nonconscious, as a good deal of cognitive science research suggests, then significant parts of our received views about morality have to be jettisoned and replaced by a new theory of moral cognition.

What we need is a cognitively realistic account of moral cognition—one that does not fall back into our inherited eighteenth-century Enlightenment conception of mind and thought. However, this is no easy task. As many ethical naturalists are painfully aware, this received Enlightenment view of moral thinking as a conscious, rational, deliberative process of bringing a particular case under some standing moral rule or principle is so deeply engrained in our received views about mind and morality that it is extremely difficult to extirpate it from our commonsense understanding. Nevertheless, we need to do so if we ever hope to have a view of moral deliberation compatible with what cognitive scientists are learning about moral reasoning.

Toward a Cognitively Realistic Conception of Moral Thinking

As we saw in chapters 1 and 2, our inherited Enlightenment conception of reason is inextricably tied to faculty psychology, by which I mean the attribution to "mind" of distinct, independent cognitive powers. Each power or "faculty" was supposed to have its own distinctive characteristics and mental functions. For example, the faculty of reason was conceived as a free, conscious, deliberative capacity to combine concepts and principles into larger coherent units of thought. Emotion was thought to be a bodily capacity to feel in certain distinct ways toward various states of affairs in the world. Imagination was alleged to be our capacity to entertain images in thought, both as they arise from sense experience and as they can be creatively combined to produce new meaning and thought. Within a faculty psychology, there can be no mixing of capacities: Reason reasons, imagination imagines, and our feeling bodily nature emotes. This rigid separation of faculties is so extreme that reason and emotion are often taken to stand in radical opposition to one another—reason works best when it avoids the influence of emotions, and emotions operate without conscious rational control.

With the rapid growth of the cognitive sciences over the past three decades, this dominant Enlightenment view of reason and emotion is coming apart at the seams, as empirical research reveals a very different picture of cognition, one in which reason is embodied, emotional, and imaginative. According to this new evidence, most of our thinking (including our moral judgment) is not a pristinely rational process in the traditional sense, and therefore reasoning is not a bloodless, emotionless, purely formal logical process. Instead, we need an intact and functioning emotional apparatus in order for our reason to have any possibility of operating appropriately in a given situation. Moreover, reasoning often involves sensory-motor areas of the brain, and therefore the shape and content of our reasoning are tied to the nature of our bodies and brains.

One very popular account that is emerging from recent experimental research on moral cognition has been dubbed by Jonathan Haidt (2001) as "the social intuitionist model." Haidt identifies a two-track process of moral judgment, one intuitive and the other rational:

Moral intuition refers to fast, automatic, and (usually) affect-laden processes in which an evaluative feeling of good-bad or like-dislike (about the actions or character of a person) appears in consciousness without any awareness of having gone through steps of search, weighing evidence, or inferring a conclusion. Moral reasoning, in contrast, is a controlled and "cooler" (less affective) process; it is conscious mental activity that consists of transforming information about people and their actions in order to reach a moral judgment or decision. (2007, 998)

The key idea here is that in a concrete situation that calls for some kind of moral assessment and decision, most of the evaluative work is done via an intuitive, emotional process in which we affectively take the measure of the situation and are moved by what Hume called sentiments of approbation or disapprobation that quickly rise up within us.[2] To be sure, our intuitive responses are typically more complex and nuanced than Hume's overly simple two-valued option of approval versus disapproval would suggest, but the upshot of these processes is the emergence of a rapidly developing motivational disposition toward what ought to be done in that situation. Our intuitive judgment comes first, followed later, if at all, by patterns of rational justification. As Haidt (2001), Marc Hauser (2006), and others have observed, a good deal of mainstream Western moral philosophy has tended to mistake the mostly after-the-fact justificatory process (in which we reflectively, consciously, and deliberately give reasons for our intuitive judgments) for the very core of our moral thinking, when in fact that justificatory reasoning is not the primary basis for our judgments and decisions.

CHAPTER THREE

As Haidt is quick to note in everything he writes on this topic, the most heated debates today are not about whether there is a two-track process (there is), but rather about what the proper weighting of the relative contribution of each process is. Is the rational process merely an afterthought that arises only in subsequent argument and attempts at justification of what we have already decided on intuitive grounds, or is there some kind of dialogical interplay between these two processes in ordinary moral thinking? In addition, if there is such an interplay, then to what extent can the rational process influence and reshape the intuitive process? Haidt (2006, 2012) likens the intuitive processes to an elephant and the "reasoning-why" processes to the rider. For the most part, the rider is along for the ride, but she can construct after-the-fact explanations or justifications for what the intuitive elephant has just done or she can make predictions about what the elephant seems about to do (46). Haidt concludes that "you can see the rider serving the elephant when people are morally dumbfounded. They have strong gut feelings about what is right and wrong, and they struggle to construct post hoc justifications for those feelings" (50). Haidt suggests that only very rarely can the rider exercise some small measure of control over what is done. He observes that "independently reasoned judgment is possible in theory but rare in practice" (2012, 46).

Another key issue is whether the "rational" process itself involves emotions and feelings. I will return to these issues later, but for now the important point is that this research suggests that we are neither what Hauser (2006) calls strictly "Kantian Creatures" nor strictly "Humean Creatures."[3] That is, we are not merely rational, principle-guided autonomous reasoners and willers (Kantians), nor are we merely emotionally driven responders (Humeans). Instead, feelings and emotions are a crucial part of our moral disposition and rational judgment, and they establish the motivational framework for all of our moral reasoning.

This two-track account of moral cognition is devastating to the traditional assumption that moral reasoning is primarily a matter of conscious analysis and application of pre-given moral principles. The empirical study of mind is leading us to look for the seat of moral judgments in a very different place than that proposed in rationalist versions of Enlightenment moral philosophy. That new place is a complex of brain regions that includes areas activated when we experience emotional responses to situations requiring problem-solving and quick decision-making. We therefore need to examine the role of emotions and feelings in our moral problem-solving.[4]

Damasio on Emotions, Life Maintenance, and Well-Being

We need a more finely textured account of the role of emotions in moral thinking. One of the best resources for such an account is the research of Antonio Damasio and his colleagues, extending over three decades of clinical and experimental work on the affective dimensions of human cognition and action. Although parts of Damasio's work are justly famous, the significance of his book *Looking for Spinoza* (2003) for providing a biologically and psychologically realistic view of moral judgment has so far not been adequately appreciated in the philosophical literature.

The parallels between Damasio's account and that of John Dewey (1922) are quite striking, although Damasio makes little mention of Dewey and does not appear to be directly influenced by Dewey's writings. As we saw earlier (chapter 2), Damasio begins his account of values where Dewey began, with the ongoing interactions of an organism (here, a human being) with its environments as the locus of all valuation. The fundamental conditions of existence of any functioning organic system are that it must (1) establish a permeable boundary that separates it from, and simultaneously relates it to, its surroundings, and (2) maintain a dynamic homeostasis (or allostasis) within that boundary.[5] *Everything else* the organism might achieve can be accomplished only if this demand for a dynamic equilibrium within the boundary of the organism is satisfied. This places homeostasis/allostasis at the heart of any acts of valuing by an organism, because it is *the* fundamental condition for the maintenance of life and well-being. Damasio argues that whatever other values might emerge will be linked to the basic life processes of the organism. To repeat a key passage quoted in the previous chapter:

I see value as indelibly tied to need, and need as tied to life. The valuations we establish in everyday social and cultural activities have a direct or indirect connection to homeostasis. . . .
Value relates directly or indirectly to survival. In the case of humans in particular, value also relates to the *quality* of that survival in the form of *well-being*. (2010, 47–48)

At the level of our *bodily* well-being, we are born with or develop quite early capacities for automatically assessing the quality of our ongoing interactions with our environment, and we then make (mostly nonconscious) adjustments, both to our bodily state and to aspects of our environment, in order to maintain or restore the equilibrium of our internal states. It is here that emotions play a central role.

CHAPTER THREE

One of Damasio's greatest theoretical contributions to date is his exploration of the role of emotions in life-sustaining and quality-enhancing dynamic homeostasis. He defines emotions as complex patterns of chemical and neural response generated by an organism's sense of its ongoing interactions with its environment. In order to sustain life and to improve the quality of that life, an organism must continually monitor and modulate its bodily states. The human internal milieu, for example, can only exist within certain fairly narrow parameters. We have to assess and adjust, without conscious effort, our temperature, hydration, salt levels, oxygen intake, integrity of the body, and scores of other conditions necessary for life maintenance and well-being. The vast majority of this body-state monitoring is done automatically and beneath the level of our conscious awareness, and thankfully so, for we could not possibly manage it all by conscious control within a real-time setting.

Damasio (1994, 1999, 2003, 2010) argues that emotional response patterns are the body's automatic ways of making the necessary adjustments to our body state (and, indirectly, to the environments that sustain that body state) so that life can go on. Brain and body mechanisms for assessing our current body state are connected with systems for inducing changes in those body states to restore and maintain bodily allostasis. Damasio concludes: "The ultimate result of the responses, directly or indirectly, is the placement of the organism in circumstances conducive to survival and well-being" (2003, 53).

Emotions are thus bodily responses to our mostly (but not exclusively) nonconscious, automatic, ongoing appraisal of how things are going for us as organisms. They help us restore disrupted allostasis, which is not merely the recovery of an earlier identical set-point equilibrium, but instead the working up, in light of changed conditions, of a new dynamic balanced state (a new allostasis) that grounds the well-being of the organism. This quest for well-being (for ourselves and others) is the very core of morality, even though it is often carried out beneath the level of our conscious acts of moral reflection.

Emotional response patterns are typically activated within our body without having to be consciously felt or reflected upon. Conscious feelings, when they do occur, are our way of being aware of the changes of our body state in response to changes in our interactions with our world.[6] A feeling, in other words, is a perception of our changing body state as it adjusts to changes in its engagement with its environment. Feelings can thus make it possible for us to be more aware of how things are going for us, and such appraisals take on special significance in situations of indeterminacy and tension where a decision about action is called for. Our conscious monitoring (as feeling) makes us better able to assess what is going on and to determine which strategies for action seem best to resolve the tensions inherent in the troubled situation: "Feelings are based on composite representations of the state of life in the process of being adjusted for survival in a state of optimal operations" (Damasio 2003, 130).[7]

Recall how Dewey described moral problem-solving as a remaking of experience in response to some awareness of a felt tension or strain in an organism-environment interaction. Damasio is likewise situating moral cognition within the life-sustaining and quality-enhancing processes of the organism's monitoring of its current body state as it engages its environment. As we will see later, this appraisal of our body state is not limited only to our own individual physical well-being, but extends also to our relations with other people in larger communities. The tentative and temporary "resolution" of a problematic situation for an organism with consciousness would thus be accompanied by a felt sense of resolution of tensions and impasses within experience. What we are feeling would be the movement from a strained to a more fluid sense of our life processes. Following Spinoza, Damasio argues that our *conatus*—our directedness, tendency, or endeavor toward certain bodily states of life maintenance and well-being in relation to the world—is experienced primarily in terms of either the forwarding or retarding of our life processes:

The fact that we, sentient and sophisticated creatures, call certain feelings positive and other feelings negative is directly related to the fluidity or strain of the life process. Fluid life states are naturally preferred by our *conatus*. We gravitate toward them. Strained life states are naturally avoided by our *conatus*. We stay away. We can sense these relationships, and we also can verify that in the trajectory of our lives fluid states that feel positive come to be associated with events that we call good, while strained life states that feel negative come to be associated with evil. (2003, 131–32)

Human flourishing is therefore experienced as positive feeling states that indicate the maintenance and well-being of the organism. This does not mean that any positive feeling state is automatically good (e.g., a drug-induced euphoria might be downright dysfunctional in a given situation), but only that our evolutionary development has laid down positive feeling states as a way of generally directing us toward well-being. However self-centered this might sound, it appears to be a basic fact of human existence, and no moral theory can be humanly adequate if it denies this grounding in an organism's survival and well-being. As we will later see, this in no way limits morality to individual self-interest, because individual well-being is inextricably intertwined with the well-being of others. As Robert Hinde (2002) and Owen Flanagan (2007) have argued, human nature has currently evolved to manifest both self-interested and other-directed dispositions.

Damasio's most radical hypothesis, in my opinion, is that maintenance of homeostasis/allostasis might be the key to human morality, insofar as emotions play the central role in our quest for survival and enhanced well-being. The key question is whether, and how well, this single notion of homeostasis can bear the full weight of moral appraisal for all aspects of the human condition. Damasio thinks it can carry most of the weight. Let us see.

Feelings as a Basis for Social Interaction

Damasio early on garnered attention for his group's work on the key role of emotions in social interactions and reasoning. In *Descartes' Error* (1994), he tells the now well-known story of Phineas Gage—who survived having a dynamite tamping rod blown up through his skull—to illustrate the range of emotional and cognitive deficits that arise from damage to certain areas of the prefrontal cortex (especially the ventromedial areas). Over the years, Damasio's team has worked with several patients suffering lesions in this area who then manifest dramatic changes in how they behave and how they interact with others. In short, people who, prior to their injury or disease, were upstanding citizens, good parents or spouses, empathic friends, and successful employees suffer any of a number of the following deficits:

- Compromised ability to make appropriate decisions in cases of uncertainty and risk
- Loss of social status
- Inability to manage their social relations
- Loss of financial independence
- Difficulty managing their daily affairs
- Impairment to their ability to plan activities
- Lack of empathy

What tends *not* to be affected by damage in these prefrontal areas is the patient's general intelligence; their ability to perceive, move, and talk;

their memory; their capacity to solve logical problems; and even their memory of conventions, rules, and moral principles that they routinely violate after their injury.

Damasio and his colleagues surmised that the areas damaged in patients exhibiting these types of impairments and dysfunction are where emotions and feelings get connected to planning and reasoning concerning actions:

> I was suggesting that when these patients faced a given situation—its options for action, and the mental representation of the outcomes of the possible actions—they failed to activate an emotion-related memory that would have helped them choose more advantageously among competing options. The patients were not making use of the emotion-related experience they had accumulated in their lifetimes. Decisions made in these emotion-impoverished circumstances led to erratic or downright negative results, especially so in terms of future consequences. (Damasio 2003, 144–45)

Some patients could literally run through in their minds quite logically a series of possible actions that might suggest themselves as options in a given problematic situation, some of which would be disastrous for their well-being, but no emotion or feeling would ever arise to direct them toward or away from any of the proposed courses of action. None of their emotionally laden past experience was available to help them choose the appropriate or most promising course of action, or to avoid socially or morally inappropriate actions. The tragic result was typically a series of bad, dysfunctional, or insensitive decisions and behaviors about their personal relationships, practical affairs, and life plans.

Damasio coined the term "somatic marker" for the gut feelings one develops through past successes and failures in the affairs of daily living. The term "somatic" picks out the role of positive or negative (pro or con) bodily feelings, while the term "marker" indicates that specific feelings get attached to certain life situations. These gut feelings guide us in determining what states and actions we ought to pursue or avoid. The positive and negative emotions that come to mark various contemplated actions do not necessarily eliminate reflective moral reasoning, but they provide the appropriate context and direct the course of the reasoning. In certain cases where they are particularly strong and well entrenched, they might short-circuit most of the reasoning process, especially in cases where our well-being is immediately threatened.

Damasio's research thus helps to explain what is going on in the processes that Haidt and others call the "intuitive" track—the track of quick,

CHAPTER THREE

automatic, emotionally charged, non-reflective appraisals and choices that drive so much of our moral cognition:

> Under the influence of social emotions (from sympathy and shame, to pride and indignation) and of those emotions that are induced by punishment and reward (variants of sorrow and joy), we gradually categorize the situations we experience—the structure of the scenarios, their components, their significance in terms of our personal narrative. Moreover, we connect the conceptual categories we form—mentally and at the related neural level—with the brain apparatus used for the triggering of emotions. For example, different options for action and different future outcomes become associated with different emotions/feelings. By virtue of those associations, when a situation that fits the profile of a certain category is revisited in our experience, we rapidly and automatically deploy the appropriate emotions. (Damasio 2003, 146–47)

These rapid, automatic, mostly unreflective emotional responses are, I take it, what is meant by intuitive moral judgments, by which we assess and choose, as it were, "before we know it." Consequently, damage to areas responsible for processing the somatic markers for various categories of experiences can lead to dramatic social dysfunction, both at the level of social interactions and moral response generally.

The Reasonableness of Affect-Driven Intuitive Processes

Is there any sense in which intuitive judgments of this sort can be called "rational"? There is. Damasio is inclined to describe these felt emotional response patterns as "reasonable" when they lead to the enhanced well-being of the organism. He notes that "in this context the term rational does not denote explicit logical reasoning but rather an association with actions or outcomes that are beneficial to the organism exhibiting emotions" (2003, 150). This is precisely the way Dewey (1938/1991, 1932/1989) uses the term "reasonable." Dewey argued that a process of deliberation is "rational" or "reasonable" by virtue of its leading to the resolution of a problematic situation, and not by virtue of its conforming to some pre-existent standard of rationality (e.g., a principle or law):

> But reasonableness is in fact a quality of an effective relationship among desires rather than a thing opposed to desire. It signifies the order, perspective, proportion which is achieved, during deliberation, out of a diversity of earlier incompatible preferences. Choice is reasonable when it induces us to act reasonably; that is, with regard to the claims of each of the competing habits and impulses. (1922, 135)

What we need to cultivate, in a sense, are reasonable emotional responses, and their reasonableness can be assessed only over the long term, as we come to see how they organize our experience. Reasonableness, then, emerges within a situation and is therefore not the agreement of an action with some preexisting rational structure or standard. Reasonableness is enacted and achieved, rather than being given and found.

The Affective Foundations of Morality

Human moral appraisals have deep affective roots. The Kantian postulation of a pure practical reason that generates actions without any intervening emotions is a chimerical vision. The only defensible interpretation of the idea that "reason is practical" amounts to little more than the claim that thinking about a proposed action and the situation it is likely to lead to activates emotional responses (pro and con) that are sufficient to move us to action. However, this is obviously not what proponents of pure practical reason intend, because they want a reason that is capable of causing, or issuing in, action without the intermediation of emotions or feelings.

In the next chapter, I will provide my account of the role of reasonable moral deliberation in our ethical appraisal and judgment. My focus in this chapter has been primarily on evidence suggesting the central role of emotions and feelings in moral cognition.[8] We now have an outline of the affective origin and sources of human moral cognition. It starts with every organism's need for life maintenance and well-being. There is nothing intrinsically "moral" about an organism's need for allostasis, although dynamic equilibrium *is* a life imperative for animals, and it is an inescapable precondition for any measure of animal well-being or flourishing. We share this life imperative with all other animals, and there is compelling evidence that some of the emotions and feelings that underlie human social behavior are present in other (non-human) species (de Waal 1996; Hauser 2006). The false presumption that morality applies to humans alone is typically predicated on the erroneous assumption that humans possess an autonomous reason and will, which is regarded as the supreme necessary condition for morality. Once we recognize how morality is present beneath and before explicit rational reflection and how emotional/feeling responses are the primary engine of our moral judgments, we are forced to abandon exclusivist claims about morality being a strictly human affair. That said, it is equally clear that human morality has achieved a complexity and richness that seems unavailable

CHAPTER THREE

to other animal species. Moreover, as we will see in the next chapter, humans have evolved modes of imaginative reasoning that are unavailable to other animals, and that allow us to explore in imagination various possible solutions to our moral problems, so it is not wrong to say that human morality is more complex, nuanced, and reflective than the moralities of other animals.

We can now ask what is it that takes us beyond mere biological regulation to full-blown human moral systems? Damasio's answer is that it is our capacity to have emotions and to experience feelings that takes us to the next level of moral engagement:

> The construction we call ethics in humans may have begun as part of an overall program of bioregulation. The embryo of ethical behaviors would have been another step in a progression that includes all the nonconscious, automated mechanisms that provide metabolic regulation; drives and motivations; emotions of diverse kinds; and feelings. Most importantly, the situations that evoke these emotions and feelings call for solutions that include cooperation. It is not difficult to imagine the emergence of justice and honor out of the practices of cooperation. (2003, 162)

Emotions and feelings, then, are the engines of our moral appraisal and action in what we are calling the "intuitive track" of moral cognition. Emotions are our on-the-ground responses to our mostly nonconscious assessment of the situations in which we find ourselves. They are part of the cycle of activity by which we monitor changes in our body state as it responds to changes in our developing situation. They generate judgment and action, for the most part, without the need for any conscious reflection or planning.

As we will see in chapter 5, there is a strong temptation among some naturalists to posit an evolutionarily developed universal human moral faculty (or instinct), as a way of explaining our capacity for intuitive moral judgments. I will argue that there is no need to posit such a faculty and that doing so is counterproductive to developing a cognitively realistic view of moral cognition. All that we need are genes that allow us to develop a number of particular perceptual, motor, and emotional response areas of our brain appropriately connected in the right ways. To cite just one example, we need to have developed those areas in the ventromedial prefrontal cortex that connect emotions with previous experiences and bring them into play in action planning and moral deliberation. This will obviously be a very complex system of cognitive and affective subsystems, but all of them could feasibly have emerged in our evolutionary history in contexts that are not intrinsically ethical, even though they eventually enable what we tend to classify as "ethical" judgments. No doubt, the tuning of these systems for moral judgment would be carried out in complex social frameworks that may well vary across cultures.

From Individual to Group Homeostasis

To summarize my previous argument, within our intuitive track, it is emotions and feelings that run the show. These feelings arise, as we have observed, as a way to monitor changes in our body state in response to changes in our surrounding environment as we interact with it. In other words, these feelings (of joy, sorrow, unease, fear, invigoration, and so forth) are the primary means by which we gauge how things are going for us, and "how things are going for us" can pertain not just to our individual well-being, but also to our broader social relations with others, and to the well-being of small- and large-scale groups or communities.

The most daunting challenge to Damasio's choice of homeostasis/allostasis as the grounding value of morality is to show how it might be at work, not just at the level of individual organism well-being, but also at the level of large groups. It is one thing to assert that life maintenance and the enhancement of an organism's well-being require continual assessment of changes in its body state (in response to changes in its environment) in order to preserve an appropriate dynamic equilibrium within the organism, but it is quite another thing to assert that allostasis is equally fundamental for our large-scale communities (Schulkin 2011). It is the awe-inspiring complexity of human interpersonal and communal life that requires us to develop additional resources for the establishment, preservation, or reconstruction of group well-being. These resources include at least the following: the full range of social emotions (such as shame, guilt, compassion, kindness, and empathy); conceptions of rights and justice; conventions and rules of behavior; public institutions to insure the smooth functioning of society; social organizations; and the arts and sciences.

Damasio connects his claims about individual homeostasis/allostasis to his hypothesis about group homeostasis as follows:

None of the institutions involved in the governance of social behavior tend to be regarded as a device to regulate life, perhaps because they often fail to do their job properly or because their immediate aims mask the connection to the life process.

CHAPTER THREE

The ultimate goal of those institutions, however, is precisely the regulation of life in a particular environment. With only slight variations of accent, on the individual or the collective, directly or indirectly, the ultimate goal of these institutions revolves around promoting life and avoiding death and enhancing well-being and reducing suffering. (2003, 166–67)

In the introduction, I articulated in a simplified manner two radically opposed options for the sources of moral values and principles. Either they descend from outside or beyond natural experience, from some allegedly supernatural or transcendent origin, or else they must emerge from the evolutionary process of our embodied, interpersonal, and cultural experience. If our values do not drop down from above, then they must work their way up from our embodied engagement with our physical/interpersonal/cultural environments. It is from this latter perspective that Damasio posits "the regulation of life in a particular environment" as the means of "promoting life and avoiding death and enhancing well-being and reducing suffering" (167). *This is the crux of morality*. There is an intimate connection between individual and communal well-being, both of which require finding a balance (i.e., a dynamic equilibrium) among competing needs, desires, interests, goals, and practices. As we have repeatedly noted, Damasio takes the key value to be homeostasis, whether within the individual or among a group of individuals:

Social conventions and ethical rules may be seen in part as expressions of the basic homeostatic arrangements at the level of society and culture. The outcome of applying the rules is the same as the outcome of basic homeostatic devices such as metabolic regulation or appetites: a balance of life to ensure survival and well-being. (168–69)

Dewey (1922) was right when he saw this life regulation as a form of problem-solving—where the problems run the gamut from basic maintenance of life processes to enhancing the quality of the organism's existence to composing a harmonious community of social creatures. What mostly separates human morality from the morality of certain non-human species is the complexity of human mind and society, especially as mediated by elaborate forms of symbolic interaction. Increasing complexity of organism-environment transactions can result in the emergence of new functional capacities of mind, thought, and language (including all forms of symbolic interaction, such as gesture, ritual, art, literature, architecture, music, and dance). The primary results of this increasing complexity are the multiple varieties of human well-being and flourishing. Flourishing is no longer merely bio-regulation, growth of the organism, and fluid action in a physical environment, but also includes many forms of individual, interpersonal, and group flourishing and meaning-making.

The dramatic consequence of this increased complexity of experience is that success in living a life of well-being can no longer be handled entirely by intuitive, automated, nonconscious, unreflective cognitive processes. *We need a more deliberative, critical, reflective track for assessing how things are going, grasping the fine textures of nuanced social interactions, proposing alternative solutions, and deciding what our best course of action might be within a problematic situation from our current perspective. We need this critical reflection because we need a way to evaluate competing values and courses of action in highly complex, indeterminate individual and social situations.* Dewey writes, "Criticism is discriminating judgment, careful appraisal, and judgment is appropriately termed criticism wherever the subject-matter of discrimination concerns goods or values" (1925/1981, 298). In addition to our intuitive emotional responses, we must be able to imagine the probable outcomes of proposed courses of action, in order to determine which course is most likely to resolve the tensions in our current problematic situation. We often need knowledge acquired from a life of experience and from scientific research on human motivation and behavior, in order to intelligently and insightfully project the way in which a current situation is likely to develop and in order to have an appropriate critical stance toward both our own perspective and motivations and those of others. As Damasio describes our condition, although there are many needs that can be quite well satisfied by automated neurochemical mechanisms, "as human societies became more complex ... human survival and well-being depended *on an additional kind of nonautomated governance* in a social and cultural space. I am referring to what we usually associate with reasoning and freedom of decision" (2003, 167; italics in original).

Dewey saw philosophy as critical reflection on the processes and methods of our valuations. Critical reflection is not an independent capacity brought externally to bear on our valuations; rather, it is a growth and development of our most basic capacity for assessment of our situation: "After the first dumb, formless experience of a thing as a good, subsequent perception of the good contains at least a germ of critical reflection. For this reason, and only for this reason, elaborate and formulated criticism is subsequently possible" (1925/1981, 300). Dewey thus posits a continuity between processes of intuitive judgment, on the one hand,

and, on the other hand, our more deliberative, reflective, critical processes of thought that are required if we are to intelligently address our most complex moral problems arising in our interpersonal and cultural contexts. Therefore, having examined the intuitive dimensions of our moral understanding and judgment, we must now ask what goes on in our more deliberate and reflective forms of moral reasoning.

FOUR

Moral Deliberation as Cognition, Imagination, and Feeling

The Third Process of Moral Cognition

We have been examining the recently popular hypothesis that our moral reasoning runs its course in two related tracks or processes, one "intuitive" and the other "rational". In other words,

> . . . intuition and conscious reasoning have different design specs. Intuitions are fast, automatic, involuntary, require little attention, appear early in development, are delivered in the absence of principled reasons, and often appear immune to counter-reasoning. Principled reasoning is slow, deliberate, thoughtful, requires considerable attention, appears late in development, justifiable, and open to carefully defended and principled counterclaims. (Hauser 2006, 31)

The claim that most of our moral appraisal and judgment is carried out by intuitive processes over which we have little, if any, rational or reflective control has led some philosophers and psychologists to drastically downplay the role of any "rational" processes in moral thinking. There is a temptation to regard reflective processes as merely after-the-fact storytelling meant mostly to explain and justify the intuitive processes that are doing the real work. For instance, even though Jonathan Haidt (2012, 67–71) allows a small and infrequent role for reflective critical reasoning in moral decision-making, he tends to narrow the scope of the rational to nothing more than rationalization of prior intuitive judgments, as when he says that "intuition (rather than passion) is the main cause of moral judgment . . . , and then reasoning typically follows that judgment . . . to construct post hoc justifications" (46). To use Haidt's favorite metaphor, the intuitive elephant is going where it will go, with its rational rider mostly along for the ride, sometimes pretending to be in control, but mostly merely reporting on and rationalizing what the elephant has done. Only on rare occasions would it even make any sense to describe the rider as in control of what the elephant does.

I agree with Hauser and Haidt that too many philosophers have focused exclusively on the "rational" processes, treating them as if they constituted the core of moral judgment, thereby completely overlooking where most of the moral evaluation and judgment actually occurs (namely, in the intuitive track). I am going to argue that there is, nonetheless, a key role for a process of moral deliberation that is more than just intuitive, nonconscious judgment, and also more than mere after-the-fact justification by principles. It is a reflective process of deliberation concerning which possible courses of action available in a given morally problematic situation would best harmonize competing impulses, values, and ends. It is an imaginative process inextricably tied to emotion and feeling, but it also makes possible an appropriately critical point of view (or what is today known as "wide reflective equilibrium"). When a process of moral deliberation achieves a sufficiently broad and comprehensive perspective, we can correctly describe the outcome (in action) of such deliberations as *reasonable*. I will then (in chapter 5) explain a reasonable judgment as a process of thinking that leads to a satisfactory resolution of a morally indeterminate situation, rather than as a process of matching our action to some allegedly pre-given rational moral imperative.

Given the current penchant for dual-process models, the question arises about where to locate moral deliberation in relation to the intuitive and rational tracks. There are two options. One can either call imaginative moral deliberation a third track (in addition to the intuitive and rational-justificatory tracks), or one can see moral deliberation as a form of *imaginative rationality*, and hence as part of a rational track (the other part of which is post hoc justification). Either option is fine. I will sometimes refer to this as a "third" process (in addition to the dual-process model), but it is important to keep in mind that this third process is indeed part of human rationality, taken in a broad sense. Human reason is not just a formal/logical structure, but rather is imaginative and affective to its very core (Johnson 1987). The key point is that there is often more going on in moral cognition than just intuitive appraisal and after-the-fact justification.

In the previous chapter, I focused primarily on the role of emotions and feelings in the intuitive processes of judgment that arise both from our bodily engagement with our environment as we seek to survive and flourish, and also from our engagement with other people in personal relationships and larger communal interactions. I will now argue that emotions and feelings also play a crucial role in the reasoning involved in our reflective deliberations, in which we assess competing alternative courses of action and come to grips with competing values. These are not just cases where a person gives after-the-fact justifications for intuitive judgments they have previously made. Rather, these processes of imaginative moral deliberation are cases of genuine ethical consideration of what one ought to do in a certain complex situation where we have the opportunity to reflectively assess our options for action.

My account of imaginative moral deliberation is the heart of my entire view of moral cognition. Because I am describing a seldom-analyzed thought process that goes significantly beyond dual-process models, my explanation has several parts and requires extended development. For the sake of the reader's sanity, I have broken this lengthy account, which is properly one long sustained treatment, into two chapters. The present chapter characterizes the nature of imaginative moral deliberation. The next chapter then explains how such a process of moral deliberation can be reasonable, appropriately non-relativistic, and subject to critical assessment.

The Need for a Reflective/Critical Process of Imaginative Moral Deliberation

The reason why intuitive, fast, unreflective moral judgments of the sort described in dual-process views are not enough to account for the full scope of our moral cognition is that they cannot provide the possibility of a sufficiently critical perspective on our intuitive judgments. It is for this reason that Dewey distinguished between what he called "direct *valuing*, in the sense of prizing and being absorbed in an object or person, and *valuation* as reflective judgment, based upon consideration of a comprehensive scheme" (Dewey and Tufts 1932/1989, 266). The intuitive processes described in chapter 3 are primarily instances of *valuing*. They are the result of the dispositions and habits of thought, desire, and action that have evolved over the course of human evolution and in response to the contingent influences that shape our individual and collective histories. Occasionally, they are even the result of prior processes of rational deliberation that have become settled habits of thought and appraisal.

The problem, however, is that, as intuitive unreflective judgments, they have no self-critical capacity. Dewey captures this unreflective and uncritical character of intuitive judgments when he writes:

> The results of prior experience, including previous conscious thinking, get taken up into direct habits, and express themselves in direct appraisals of value. Most of our judgments are intuitive, but this fact is not a proof of the existence of a separate faculty of moral insight, but is the result of past experience funded into direct outlook upon the scene of life. . . . There is a permanent limit to the value of even the best of the intuitive appraisals of which we have been speaking. These are dependable in the degree in which conditions and objects of esteem are fairly uniform and recurrent. . . . Taken in and of themselves, intuitions or immediate feelings of what is good and bad are of psychological rather than moral import. They are indications of formed habits rather than adequate evidence of what should be approved and disapproved. They afford at most, when habits already existing are of good character, a *presumption* of correctness, and are guides, clews. (Dewey and Tufts 1932/1989, 266–67)

In chapters 2 and 3, I gave my account of where these intuitively prized values come from (i.e., what their sources are in our biological, interpersonal, and cultural makeup) and how they dispose us to pursue certain states of affairs and avoid others. However, there are at least two important reasons why we should not rest content with only the intuitive track of appraisal and judgment. First, what we regard as our intuitively given values have emerged in response to certain types of situations and conditions, but when any new, unfamiliar conditions arise and new problems emerge, our received values and standards may not be adequate for dealing with the changed conditions we are encountering. Our entrenched dispositions and values may not have been "designed" (i.e., may not have emerged evolutionarily and historically) with those new, unanticipated conditions in view. Consequently, we cannot simply assume that our intuitive, non-reflective appraisals and valuings are adequate when we encounter new conditions and complexities.

The second problem is that the fact that we have come to value certain states of affairs is no guarantee that we *should* value them in the way we do. Racial prejudice, gender discrimination, and religious intolerance are deeply and profoundly impressed on the heart and mind of most of us. The strength of conviction we might have in expressing any of these prejudices (as habitual pre-judgments) is no sign of their moral correctness, but only of their smothering hold on us.

What we most need, therefore, is some way to achieve an appropriate critical distance on our *valuings*, that is, on our acquired habits and value judgments. We need a process that Dewey called "valuation," which he saw as requiring "conscientiousness" in our reflective moral deliberations, in pursuit of a broader critical perspective:

> Perhaps the most striking difference between immediate sensitiveness, or "intuition," and "conscientiousness" as reflective interest, is that the former tends to rest upon the plane of achieved goods, while the latter is on the outlook for something *better*. The truly conscientious person not only uses a standard in judging, but is concerned to revise and improve his standard. (Dewey and Tufts 1932/1989, 273)

Intuitive appraisals tend to be based on what we earlier called "homeostatic set points" (values) to which the organism (here, a human being) seeks to return. In contrast, because reflective deliberation is a response to changing conditions, it must seek allostasis, or the establishing of a *new dynamic equilibrium that is responsive to the altered conditions*. What we most need in such cases of novel circumstances is conscientious imaginative deliberation.

A Deweyan View of Moral Deliberation as Reflective Valuation

In a series of groundbreaking books, including *Human Nature and Conduct* (1922), *Experience and Nature* (1925), *Ethics* (1932), and *Logic: The Theory of Inquiry* (1938), Dewey proposed a theory of human inquiry in which reasoning actually *transforms* the character of experience, remaking and remolding it, instead of merely reporting on it after the fact. Reasoning is a cognitive-conative-affective process of action, and it incorporates qualities, feelings, and values at its very core (Johnson 1987, 2007).

One of my main theses in this book is that moral deliberation is a complex form of human problem-solving that emerges from our embodied, visceral engagement with our world and then reconstructs that world through a projective imaginative process that can be both affect-laden and reasonable. In chapter 1, I quoted Dewey's well-known account, in *Human Nature and Conduct*, of moral deliberation as a process of dramatic imaginative rehearsal of possible courses of action, in order to determine which available course best resolves the problematic situation we find ourselves in. A decade later, in *Ethics* (1932), Dewey offers a closely parallel summary statement about the nature of deliberation:

> Deliberation is actually an imaginative rehearsal of various courses of conduct. We give way, *in our mind*, to some impulses; we try, *in our mind*, some plan. Following its career through various steps, we find ourselves in imagination in the presence of the consequences that would follow; and as we then like and approve, or dislike and disapprove, these consequences, we find the original impulse or plan good or bad. Deliberation is dramatic and active, not mathematical and impersonal; and hence it has the intuitive, the direct factor in it. (Dewey and Tufts 1932/1989, 275; italics in original)

Moral deliberation is an imaginative process of inquiry that arises when changes in our situation call into question our ordinary habits of thought, feeling, appraisal, and action—that is, our intuitive valuings inherited from prior experience. Because our routine habits of valuing are blocked or frustrated, we find ourselves in a problematic situation characterized by indeterminacy and tension about how we should go forward. Moral inquiry is the process by which we come, first, to define the problem at hand, and then explore in imagination how best to resolve the tension in the present situation.

This Deweyan conception of moral deliberation is so different from the accounts provided by many classic moral theories that it merits a deeper exploration of its inner workings. What follows is an account of the nature of the major dimensions of this reflective process.

1. The Specificity of the Lived Situation

As Dewey repeatedly observed, one of the great mistakes of so many moral theories is their failure to start from *experience as it is lived*. Instead, they begin by assuming that their theoretical abstractions by which they try to make sense of moral cognition are actually constitutive of the moral experience of ordinary folks. The error is to read back into our experience abstractions like "ends," "pleasure," "principles," "virtues," "goods," "duties," "rights," and "happiness," and then to overlook any part of experience that does not fall under such categories of thought. Although sometimes relevant to our moral reflections, such abstractions oversimplify by directing us away from the complexity and richness of our actual lived situation. Dewey defined a *situation* as "a complex existence that is held together in spite of its internal complexity by the fact that it is dominated and characterized throughout by a single quality"

(Dewey 1930a/1988, 246).[1] What always comes first and contextualizes everything else that we think, feel, or do is the enveloping situation we find ourselves in at a particular moment in time. If it is a situation of moral indeterminacy and tension, then we need to understand that situation in all its complexity, if we ever hope to transform it for the better.

It is only within this environing experiential horizon that objects, events, and actions stand out for us and have the meaning they do. We do not experience isolated objects, isolated individual qualities, or isolated discrete actions.[2] Objects and actions emerge in context and have their meaning via that particular context. We discriminate (often automatically and without conscious control or intent) what seems important to us in that particular situation, at that particular moment. What matters to us will depend partly on our biological makeup, partly on our interpersonal relations, and partly on our cultural institutions and practices (Johnson 2007; Eagleton 2011). Consequently, any objects, events, and actions we deal with provide what J. J. Gibson (1979) called "affordances"—possibilities for experience constituted by the interaction of our perceptual and motor capacities with the features of our surrounding environment. For example, cups afford graspability for humans, but not for snails, which encounter cups as climb-up-able. Affordances thus require an active organism engaging aspects of its environment that the organism has evolved to notice and interact with. To have a concept of a certain object or event is to be able to enact a simulation of the experiential possibilities (past, present, and future) for interacting with that object. It is within this world of qualitative affordances that we act to discern, inquire into, and realize certain qualities. However, the affordances provided by any object, person, or event are not based only on our sensory and motor capacities, but equally on our makeup as social beings with cultural habits and values. Therefore, I would extend Gibson's notion of affordances beyond physical objects and environments to include possibilities for social interactions in a cultural context. We could thus say that a certain interpersonal or social situation affords, for creatures like us with our cultural upbringing, a certain range of possible modes of response and action.

For moral theory, the significance of the primacy of the concrete situation is paramount. One important consequence of this is that moral problems are specific and context-dependent. They are not, in the first instance, generic in their character. The problems we have to deal with are *our* problems, *here and now*. As such, they bring with them all of the complexity, richness, and specificity of our current situation, and the failure to appreciate this complexity leads to "solutions" that are not good solutions because they do not actually address the specific circumstances we are facing. No doubt, what turns out to be "best" for one particular situation will sometimes have broader significance and implications for other situations, insofar as they are similar in their basic characteristics. However, the situation we must identify and evaluate is the situation as we encounter it *now*, in all its specificity. Gregory Pappas sums this up when he observes, "For Dewey, radical empiricism in ethics entails a radical contextualism, by which he meant that each situation constitutes a unique context and while it is lived (as a process), that is all there is to moral life" (2008, 41).

Moral theories too often succumb to the temptation to treat situations abstractly (as kinds of problems), and then they look for universal ethical principles governing those generic situations, thereby overlooking what is actually going on in the situation at hand. One sees the situations through his or her preferred abstractions. Pappas examines some of the many ways that theories of morality can go astray by failing to start with the experiential situation in its fullest, richest, most complex sense. Some theories begin by assuming that morality is primarily about acts of judgment under universal principles; others claim that morality stems from emotions alone; others treat moral agents as, by nature, self-interested satisfaction maximizers; and still others regard morality as being about realizing pre-given fully determinate ends. Although such ways of parsing a situation may have some value, the problem arises whenever a theorist reads assumptions like these about human cognition and behavior back into experience, as if their particular abstractions captured the core of our moral experience.[3] The point is that selectively abstracting these and other dimensions, and then pretending that they fully define the relevant moral considerations, leads us to overlook the complexity of human experience within particular situations. Such a limited viewpoint can never do justice to the conflicting habits, interests, practices, and values that are the locus of our moral problems. Pappas captures this point when he concludes that "moral theorists have the tendency to forget the non-cognitive complexity, plurality, incommensurability, raggedness, changeability, and uniqueness that characterize our primary moral experience. These are the features of moral life that almost make them recalcitrant to any kind of theoretical formulation" (2008, 45).

2. Moral Deliberation Is Tied to the Qualitative Unity of the Situation

If good moral inquiry must begin with the concrete fullness of our specific situation, then how do we know what the true character of any particular situation is? Dewey's somewhat disconcerting answer is that every situation is unified and given its unique character by a pervasive quality. In other words, the specificity of the situation is realized as a qualitative unity. Consequently, moral deliberation must be a fundamentally *qualitative* process, arising from our qualitative world and enacting change within that world. Dewey insisted that, unless our initial deliberations are rooted in the qualitative character of our circumstances, our moral thinking will never be grounded in and responsive to our actual, lived, concrete moral situation. Dewey introduces this striking notion as follows:

The world in which we immediately live, that in which we strive, succeed, and are defeated is preeminently a qualitative world. What we act for, suffer, and enjoy are things in their qualitative determinations. The world forms the field of characteristic modes of thinking, characteristic in that thought is definitely regulated by qualitative considerations. (1930a/1988, 243)

Grounded thought and action take place in relation to the qualities that make up our current situation—the pitch and timbre of a baby's cry, the texture of our lover's skin, the fear in our children's eyes, the thick dark redness of blood pouring from a wound, the quality of light on a late winter afternoon, the anxiety in a coworker's voice, the roughness of the path under our feet, the slightly tart sweetness of fresh raspberries, the conflict in our heated conversation. At our most visceral level of functioning, the meaning, values, thoughts, and plans that engage us are primarily influenced by the qualities available to us through our interaction with our environments. These qualities are not just perceptual properties of objects, but include qualities of our felt relations with other people and our cultural practices and values. We seek, for instance, to ease the tension and hostility evident in an office discussion of politics. Our sense that we have managed to reduce tensions or harmonize apparently conflicting ends is as much felt as it is thought.

One of Dewey's signature claims is that any developed situation will be integrated and organized by what he calls a "pervasive unifying quality." Every situation is individuated by a unifying quality that colors and cements its parts into a unified whole with a distinctive character. Dewey often used artworks to illustrate the notion of a pervasive quality, because he regarded the arts as exemplary enactments of the qualitative dimensions of situations and as consummations of meaning. For example, a Bach fugue has a fundamentally different qualitative unity than does a Brahms quartet, just as a blue-period Picasso makes a different overall qualitative impression than does a Matisse paper cutout or a Warhol silkscreen print of Chairman Mao. You can hear the "Bach-ness" in a Bach fugue, just as you can hear the "Brahms-ness" in a Brahms quartet.

However, the demarcating qualitative unity is not just a vague felt generic impression (like *Bach-ness*) that lets us distinguish styles and artists. Rather, *every particular work and every particular situation has its own distinguishing pervasive quality* that separates it from other artworks and other situations.[4] For instance, Bach's "gigue" fugue has a different unifying quality than his *Toccata and Fugue in D Minor*. Van Gogh's *Starry Night* has a different qualitative unity than does his painting of his bedroom at Arles. The point here is that situations—artistic, moral, or otherwise—are each marked off and individuated by some particular qualitative unity. It is in relation to the felt unifying quality of a particular situation that we are able to selectively abstract objects, qualities, and relations that we act in response to.

Viewed from the perspective of a person experiencing a particular situation, we are inclined to say that they subjectively "feel" or "experience" its unifying, individuating quality. Although it is true that they *do* feel it, that does not make the quality merely subjective or personal. The qualitative unity is *in* and *of* the situation. In that sense, it is objective, shared, and available to others. The situation in which a person is kidnapped by terrorists is terrifying, and the terror felt by the kidnap victim is the subjective dimension of the objective terror of the situation. The moral anguish of an unmarried teenage woman who finds herself pregnant is not merely *her* anguish; rather, her situation is anguished, conflicted, indeterminate, and terrifying. Your argument with a close friend over whether to join the faculty union is tense, strained, indeterminate, and consequential for the future of your life, your relationship with your friend, your moral identity, and perhaps even the future of the union. This is not just a matter concerning your feelings about how things are going. The *situation* is strained and tense, the *situation* is problematic, the outcome of your conversation is indeterminate, and the future of your relationship is uncertain. Moreover, it is the distinctive problematic, indeterminate, uncertain character of the situation that makes it *that* particular situation that you need to deal with and respond to at a qualitative level.

Dewey's conception of a situation has significant implications for the nature of moral reasoning. Most importantly, moral deliberation is a form of problem-solving geared toward the transformation of a situation that is problematic and indeterminate into a situation that allows an individual or group to move forward in a more or less satisfactory way. *The* "test" of any process of moral deliberation is how it resolves the problematic character of that particular situation by changing its pervasive unifying quality. The key to efficacious moral deliberation is to grasp the problematic situation in all its breadth, depth, complexity, and richness, and then to respond intelligently to that situation as a way of reconstructing it for the better.

The most troubling component of this account is to explain what the pervasive unifying quality of a situation has to do with right action. The problem, in other words, is to figure out what Dewey meant, in the quotation just cited, when he said, "The world forms the field of characteristic modes of thinking, characteristic in that thought is definitely regulated by qualitative considerations" (1930a/1988, 243). *Thought is regulated by qualitative considerations!* That, in my opinion, was Dewey's most radical claim—that all our perceiving, valuing, judging, knowing, and doing get their meaning, force, and direction from the total qualitatively unified situation we find ourselves inhabiting at a given moment in time. Relevant and appropriately constructive moral reasoning is inextricably rooted in and shaped by the pervasive unifying qualities of particular situations:

As a matter of fact, such intellectual definiteness and coherence as the objects and criticisms of esthetic and moral subjects possess is due to their being controlled by the quality of subject-matter as a whole. Consideration of the meaning of regulation by an underlying and pervasive quality is the theme of this article. . . . The special point made is that the selective determination and relation of objects in thought is controlled by reference to a situation—to that which is constituted by a pervasive and internally integrating quality, so that failure to acknowledge the situation leaves, in the end, the logical force of objects and their relations inexplicable. (Dewey 1930a/1988, 246)

The primacy of the situation is another way of stressing the primacy of experience for any form of inquiry. What could seem more obvious than that moral deliberation must begin and end with experience—experience not just as a subjective response, but rather as a process of organism-environment interaction, and so as objective? And yet this is a radical hypothesis, one that challenges a long history of philosophical neglect of experience in its deepest, most complex lived sense. Dewey devoted large parts of his massive corpus of writings to showing the dangerous consequences of our philosophical tendency to oversimplify and selectively abstract from experience, and then to proceed as if our abstractions captured everything that could matter in experience!

I will argue later (chapter 7) that the greatest sin of impoverished moral understanding is the repeated failure to appreciate the complexity of our moral situations, which leads to the corresponding mistake of drastically oversimplifying the matters at hand (as in, "The fetus is a person, or it is not" and "Abortion is murder, or it is not"). The result of this failure to grasp the full situation is our tendency to settle for partial, highly selective, and often reductive accounts of moral experience and deliberation. This failure is enacted in our actual moral deliberation as the tendency to ignore some of the relevant factors that define our situation. The most common selective abstractions are moral rules or imperatives, but other reductive abstractions can focus exclusively on any of the following: feelings, sentiments, self-interest, will, pleasure, ends, freedom, intuitions, alleged moral instincts, or postulated moral facts.

It is not hard to see why someone harboring a traditional conception of logic—as a preexisting, transcendentally grounded structure of possible formal relations—would be dismissive of Dewey's insistence on qualitative dimensions. Dewey's is a *logic of inquiry*, that is, an account of logical principles and norms arising from forms of engaged practical inquiry that have proved themselves valuable in our prior attempts to deal constructively with the actual problems that we have confronted. Instead of being absolute givens of transcendent formal relationships, logical principles are patterns useful for conducting various types of thinking, inquiry, and problem-solving. Logic and its norms are not *brought to* experience from outside, but rather *arise from* experience and are appropriated as tools for reworking experience by means of intelligent inquiry. From such a perspective, moral principles are never absolute laws, but are instead tentative guides to reasonable processes of inquiry.

The great problem with making good on the promise of a situational or experientially based account of moral inquiry is that almost nothing can be said, in a general way, about how the qualitative unity of a situation determines for us what is important and guides the direction of our thinking (Gendlin 1997). This problem has been recognized by many philosophers who question whether the notion of a "pervasive unifying quality" does any actual work. Richard Shusterman, for example, has a broadly Deweyan perspective on aesthetics, ethics, and inquiry, but even he expresses a deep concern about how reference to the defining quality of a situation could actually generate norms. He asks, "How are we supposed to measure (let alone communicate) magnitudes of an experience which cannot even properly be defined or marked off for measurement? There is no way the critic can prove his verdict to others by mere appeal to his immediate experience, which in its qualitative immediacy and tremulous transiency is not discursively demonstrable or preservable as evidence" (2000, 56). Shusterman is here referring to aesthetic judgment, but his concern applies equally to Dewey's account of moral deliberation as grounded in the qualitative situation. From a different perspective, Colin Koopman (2009), following Rorty, has also challenged reference to primary qualitative experience, raising questions about whether it verges on falling back into another type of foundationalism about values. He fears that the felt unity of the situation is being used as a new kind of given, in order to ground some form of judgment and knowing.

The root of the problem here is that, as soon as we begin to conceptualize which aspects of a situation determine relevance and how they do this, we are already abstracting out and focusing on particular objects, relations, and qualities, thereby missing the pervasive unifying quality that characterizes the whole situation. As Dewey puts it, the situation "forms the universe of discourse of whatever is expressly stated or of what appears as a term in a proposition. The situation cannot present itself as an element in a proposition any more than a universe of discourse can appear as a member of discourse within that universe" (1930a/1988, 247).

In other words, our thinking can only be appropriate to our situation when it operates on materials that are relevant to the problem at hand, and that is something determined by the total situation, not by abstractions from it. Even if you conceive of morality as a system of rational principles, you would have to somehow determine which aspects of a given situation are relevant in deciding which principle applies and how it applies. Dewey's claim is that, insofar as our reasoning actually does solve our current problem, it must have discovered the relevant details of the situation at hand, and this must be determined by the unifying quality of the situation:

The underlying unity of qualitativeness regulates pertinence or relevancy and force of every distinction and relation; it guides selection and rejection and the manner of utilization of all explicit terms. This quality enables us to keep thinking about one problem without our having constantly to stop to ask ourselves what it is after all that we are thinking about. We are aware of it not by itself but as the background, the thread, and the directive clue in what we do expressly think of. For the latter things are *its* distinctions and relations. (Dewey 1930a/1988, 247–48)

What this entails for moral deliberation is that any moral reflection depends on a prior felt sense of what the important elements, relations, and qualities are within a particular problematic situation. This is what is meant by having an "intuitive sense" or "grasp" of a situation. Dewey explains: "The problem is had or experienced before it can be stated or set forth; but it is had as an immediate quality of the whole situation. The sense of something problematic, of something perplexing and to be resolved, marks the presence of something pervading all elements and considerations. Thought is the operation by which it is converted into pertinent and coherent terms" (1930a/1988, 249).

As we have already observed, our intuitive take on a situation arises from past experience in which we have developed sedimented patterns of emotional response to various "emotionally competent stimuli" (Damasio 2003). The character of these particular affective responses will depend on biological makeup, cultural factors, and personal history; that is, our responses will depend on how we have evolved to experience certain kinds of emotions in certain types of situations, how our cultural practices and values have shaped our emotional responses, and how our individual past experience has determined some of the particularities of our tendency to respond in certain ways to certain types of situations.

In the previous chapter, we saw that emotions are complex patterns of bodily response generated by our mostly nonconscious monitoring of our body states, insofar as that monitoring reveals "how things are going for us." Emotions are our most visceral way of assessing our well-being and of trying to further that well-being (and the well-being of others) through changes in our body state and our environment. Appropriate emotional responses to our situation are those that are evoked by the unifying character of our situation and that allow us to transform our situation for the better. This is an *affective* process of qualitative change that typically has little to do with the central claim of Moral Law folk theory that there are moral laws given *a priori* and that moral reasoning is a problem of determining if and how situations fall under those laws or rules. Instead, the process is better described as affectively taking the measure of our situation and reworking it to resolve the blockages and tensions we discern within the situation.

The big question, obviously, is how we come to know that a problematic situation is to some degree resolved. The answer can only be that *we feel or experience the transformation of the situation from one that is problematic, indeterminate, and conflicted to a situation that is clarified, harmonized, and unified in a way that carries us forward in life.* Could this possibly be an accurate description of moral deliberation? To determine if this is a workable account, we need to investigate carefully the entire process, from the sense of a problem, to the defining of the problem, to the exploration of possible solutions, to the choice of one as better in the present situation. This investigation will require the remainder of the present chapter and all of the next.

3. The First Stage of Moral Inquiry: Defining the Problem to Be Addressed

Many moral theories mistakenly assume that we typically encounter a clearly defined moral issue, and then we must find the right moral standard or principle under which it falls. Dewey observed that often the relevant character of our situation is initially *not* clearly defined, and so it is a crucial stage of our deliberation to define the problem in the right way. Moral reasoning is a multistage process of experiential reconfiguration. It begins with a situation in which our prior habits of thought and action are experienced as inadequate for dealing with new conditions that have arisen within our situation. This sense of inadequacy is felt as a tension, conflict, or frustration. Your habitual modes of response and action are not fitted to the changing circumstances, and it is not clear how you should proceed.

Upon encountering such a blockage, you have two options—either (1) you can strap on the blinders and plow ahead with "business as usual," trusting that your prior habits, values, and principles are actually all that is needed to see you through, or else (2) you can acknowledge the need for moral inquiry that takes you beyond your preestablished values and patterns of inquiry. The first path assumes that the right standards must be pre-given, if only we could get clear about how they give us correct guidance in the present case. The second path recognizes that changed conditions appear to require a reconsideration of our habitual modes of thought and judgment.

Whenever we encounter a moral difficulty, the temporary blockage of prior habits of thought and action impedes further action and leads to confusion. However, this pause in the smooth habitual flow of our lives can be a crucial occasion for just the kind of reflective moral inquiry we need, given that something we have been doing is no longer working. The inquiry that emerges is a form of action capable of transforming our situation when prior habits have failed us.

A key part of this reconstructive process involves coming to some awareness and understanding of the *nature of the problem* one is facing. This critical move is essential, because the nature of the problem is *not just given*; it has to be discovered through inquiry. Identifying and defining your problem is often the most important stage of your moral inquiry, insofar as failure to correctly define the problem makes it impossible to forge solutions that are responsive to the actual situation you find yourself in.

Consider, for example, the issue of the morality of drone warfare.[5] From one perspective, the "problem" can be circumscribed as essentially a concern about national security. Assuming that there are genuine terrorist threats against a sovereign nation, some of which have been carried out, the president reasons that drone warfare is the best option we have for combating terrorism. First, as compared with aerial bombing strikes, drones tend to inflict less "collateral" harm on civilian populations. Second, compared to "boots on the ground" combat operations, they do not risk the combatant casualties of employing ground forces. Third, they are far less expensive than other options, which is a consideration for any country with a massive national defense budget.

However, there are other ways one might define the problematic situation and circumscribe the relevant moral considerations. One much-debated issue is the possible killing of U.S. citizens without due process of law. A sovereign leader takes the advice of military intelligence that some individual, perhaps an American citizen, poses a threat to our security. A drone strike is ordered, but this amounts to killing a citizen without due process of law, since this person has neither been formally charged with a crime, tried by law, or convicted of a crime.

A third, slightly less tangible, consideration might be the nature of drone warfare itself, in which there is nearly a total removal of the perpetrators of the action from the awful human effects of their deeds. In a war room full of computers and infrared site monitors, the decision is made to push a button. Beyond a flash on the screen, there is little visceral connection to what has just been done.

The point here is that the way one circumscribes the relevant phenomena that come to define the situation will make all the difference in what one ought, or ought not, to do. Someone might counter that the nature of the problem is evident, because, obviously, you have to consider *all* possible perspectives (to get the full picture) before deciding what to do. I agree, but this does not solve the problem of how we determine when we have "the full picture," nor does it give us a way to determine which considerations are morally relevant. Even if we assume that we need to seek the most comprehensive understanding of the complexity of the moral problem, what this perspective is will not be a given "fact of the matter." Therefore, defining the problem is a crucial stage in the moral inquiry, not a "given" from which we start our deliberations.

Most of the time, the problems we are facing do not wear their identity on their sleeve. Quite often, when we have a genuine moral difficulty, the character of the problem is not obvious; rather, articulating a problem is already half of the deliberative process, and this requires considerable reflection. Identifying a problem is thus a major stage in the inquiry, because without this stage, any deliberation is likely to be relatively unresponsive to the complex situation you are actually facing, leaving you unable to explore relevant solutions and therefore unable to resolve the tensions in a reasonable manner. *The nature of the problem is arrived at through a process of reflection and inquiry*, which is itself a transformation of the situation from an initial state of helpless indeterminacy into a new situation that is determinate enough (because we have identified a problem) so that *relevant* possible solutions can now be explored. Prior to your determination of the nature of the problem, there is no intelligent way to discern the range of available future courses of action that would be appropriately responsive to the qualities of the situation you find yourself entangled in. The ability to sensitively and insightfully identify the fullest range of competing values, claims, principles, motivations, and habits is a deliberative skill that can be performed well or poorly.

It is only at this stage of moral deliberation that we can begin to imaginatively envision each available course of action in turn, until we settle on one that permits us to overcome the blockage of our former habits and carry our experience forward in some measure. Deliberation is not a precondition of, nor a preparation for, action. Rather, *deliberation is itself a phase of action*. It is a process that achieves its tentative completion in some action that reconfigures our situation. Furthermore, since the institution of *new* habits of behavior is a reformation of one's character and dispositions to act, moral deliberation reconstitutes the self (chapter 8). Finally, on the issue of identifying a problem for moral consideration, Colin Koopman has expressed the worry that, although Dewey has major insights about methods of inquiry and problem-solving, he has little to offer concerning how we might come to problematize a situation: "Although Dewey devoted hundreds of pages to the genesis of solutions, he hardly ever wrote about the genesis of problems. Dewey simply had too little to say about how we might fashion forms of inquiry that help us bring the right kinds of problems into focus" (2009, 199). Dewey gives a central role to our experience of an indeterminate (problematic) state of affairs as the first stage of moral inquiry and the impetus for further moral inquiry, but Koopman asks, "What about determinate situations that are in need of indeterminacy? What about situations where no conflict is felt or perceived by the dominant social group? What about racism? How do we problematize (make-indeterminate) what already 'feels' so right?" (Koopman, personal communication, 8/1/12; and also Koopman 2009).

Koopman offers Michel Foucault's genealogies of power, subjectivity, and institutional disciplinary practices as exemplary ways of revealing how our identities are often situated within structures of power relations that are beyond the control of any individual. Foucault's archaeologies of knowledge and power give us a way of problematizing situations where those in power, and often those who are subject to these power relations, may see no problem at all.

I see no reason to think that Dewey would not have welcomed Foucault-type analyses as an important strategy for uncovering ethical problems. In fact, Dewey's extensive writings are full of trenchant analyses of historical practices, institutions, and events that have fatefully determined our current values, the methods of inquiry we employ, our conceptual systems, and our habits of thought and action. Dewey was no slouch as a genealogist of knowledge and practices, and he is often a master at calling into question what has been taken for granted in various philosophical traditions.[6]

Besides sophisticated and nuanced Foucaultian mining of our institutional and cultural practices, there are other ways to unsettle the complacent determinacies that constitute cultural blind spots. History is not lacking in prophets, by which I mean not only classical biblical prophets like Elijah or John the Baptist, or Greek seers like Tiresias, but prophets and visionaries like Martin Luther King Jr. and Nelson Mandela. They call us to account for our sins of commission and omission. They demand conscientiousness about moral matters we do not wish to behold, and they challenge us to change our ways.

On a less grand scale, there are other voices of conscience—ordinary folks who simply "cannot take it anymore," and so speak out, by words and actions, against whatever injustice, inequality, or degradation they can no longer tolerate in silence. I am thinking here of any of thousands of people who see problems and go out on a limb to bring them before our individual, national, and global consciousness. Jane Addams, Karl Marx, unionists, Gandhi, peace activists, Pete Seeger, the mature Bobby Kennedy, Aung San Suu Kyi, many Occupy participants, and on and on, throughout human history.

In short, Koopman is right when he argues that we need to enrich our moral problem-solving to encompass forms of inquiry for unsettling the settled, problematizing the apparently unproblematic, and expanding our habits of intellectual discernment. Koopman's (2009) and John Stuhr's (1997) "genealogical pragmatisms" remind us that we need to find ways to open ourselves to other perspectives that might challenge what we so confidently take for granted. However, I would argue that Dewey was actually one of the better practitioners of this type of genealogical inquiry into norms and assumptions underlying our inherited conceptual frameworks.

4. Dramatic Rehearsal as Cognitive-Conative-Affective Simulation

In moral deliberation, if the end were always well defined, determinate, and pre-given, then our task would be reduced to nothing more than means-ends reasoning, in which we find the most efficient means to the realization of a preestablished end. In truly problematic cases, on the contrary, the ends themselves are in question, and therefore a crucial part of the deliberative process is working toward an emergent end that becomes clarified as the process of inquiry develops. Moral deliberation requires us to run out in imagination various alternative courses of action available to us, in order to see which course best resolves the indeterminacy in our situation. The process required is what today cognitive scientists call "mental simulation."

Lawrence Barsalou (1999, 2003) has developed the beginnings of a "simulation semantics" as a radical alternative to classical theories of conceptualization and category structure.[7] According to classical theories of categorization, our conceptual categories are formed as perceptual inputs, organized by concepts, and stored in semantic memory for later recall and redeployment.

In contrast, *simulation semantics treats conceptualization and reasoning as experiential simulations:* "A concept is not a single abstracted representation for a category, but is instead a skill for constructing idiosyncratic representations tailored to the current needs of the situated action" (Barsalou 2003, 521). The constructed representations are not lists of defining features of objects, nor are they abstract conceptual entities, but rather involve perceptual and motor simulations of what it is like to experience and interact with certain kinds of objects, people, and situations. For example, to have a concept *car* involves, among other things, capacities to simulate what cars look like and how we physically interact with them.[8] The perceptual simulations would be of views of cars from multiple perspectives, focusing successfully on different parts of the car (side view, doors, tires, windows, seats, dashboard). The motor simulations would capture the kinds of bodily interactions we can have with cars, such as opening doors, sitting in seats, manipulating controls, driving, and so on. Conceptual systems emerge within a context of action in the world, and therefore concepts are structured in relation to situated perception and action. Consequently, what we call concepts will include simulations that occur in visual, tactile, olfactory, aural, and proprioceptive modalities, as well as various motor programs for interacting with the particular object or situation marked out by the concept. Conceptual simulations will evoke the emotions and feeling dimensions of the experiences being simulated. They will also include social, cultural, aesthetic, economic, and religious dimensions of our experience of cars. Since concepts are dynamical systems, they can change over time as the result of changes in our bodies and the physical and cultural environments we interact with.

Given the dynamic, action-oriented nature of conceptual systems, one would understand problem-solving and reasoning as modes of multimodal action. Barsalou defines problem-solving as "the process of constructing a perceptual simulation that leads from an initial state to a goal state" (1999, 605). This is too narrow, insofar as is assumes that the goal state (end) will always be given, but it points the way to a broader, more adequate notion. As just noted, we want to avoid the reduction of moral deliberation to means-end reasoning alone, as if the end were given and the means thereto were then calculated. Instead, oftentimes the end is not fixed and pre-given, but can only be made determinate and clarified as the result of some process of deliberative simulation. Although we may begin our deliberation with some ends-in-view, those ends can be, and often are, redefined through the deliberative process, as inquiry investigates the complexities and nuances of experience. Barsalou comes closer to Dewey's conception of deliberation when he defines decision-making as "specializing a simulated plan in different ways to see which specialization produces the best outcome.... Skill results from compiling simulations for most of the plans in a domain through extensive experience" (1999, 605). "Best outcome" here is not to be defined merely as maximizing of satisfactions (though that may be part of it), but rather refers to a far more comprehensive assessment of how we can best go forward in that situation so defined.

Ben Bergen has also proposed a "simulation semantics" supported by scores of recent experimental studies of some of the ways our understanding of language involves simulations of perceptions and actions. Bergen's version of simulation semantics argues for the "embodied simulation hypothesis" that "we understand language by simulating in our minds what it would be like to experience the things that the language describes" (2012, 13). Our processing of meaning in language involves enacting patterns of perception, feeling, and bodily activity that we have come to associate with some object, event, or state of affairs:

Simulation is the creation of mental experiences of perception and action, in the absence of their external manifestation. That is, it's having the experience of seeing without the sights actually being there, or having the experience of performing an action without actually moving. When we're consciously aware of them, these simulation experiences feel qualitatively like actual perception; colors appear as they appear when directly perceived, and actions feel like they feel when we perform them. The theory proposes that embodied simulation makes use of the same parts of the brain that are dedicated to directly interacting with the world. When we simulate seeing, we use the parts of the brain that allow us to see the world; when we imagine performing actions, the parts of the brain that direct physical action light up. The idea is that simulation creates echoes in our brains of previous experiences, attenuated resonances of brain patterns that were active during previous perceptual and motor experiences. (Bergen 2012, 14–15)

The bulk of Bergen's book reports experiments that reveal many of the ways in which reading or hearing sentences activates simulations of the events described, in a way that uses the parts of our brain that would be activated if we were actually to have those experiences or perform those actions presented in the sentences. Visual descriptions activate regions of the visual cortex (and other subcortical regions) that would be involved in our actually seeing what is described. Described motions activate areas of the brain responsible for perceiving motion and for making the appropriate motion with one's body. When we process sentences describing actions that utilize body parts (e.g., "she *gripped* the ball," "Jane *threw* out the first pitch," "he *kicked* the football"), we activate somatosensory areas related to the various appropriate body parts, such as hands, legs, and feet.

As we saw in chapter 2, studies of mirror neurons in monkeys and humans indicate that the same brain areas are activated when we see someone performing a specific action as when we perform that precise action (Gallese et al. 1996). This appears to hold true for imagined scenes, actions, and experiences, although fewer studies have been done on this (Decety and Grezes 2006; Aziz-Zadeh et al. 2006). Other experiments suggest that these processes of simulation could be the basis for our capacity to feel empathy (Decety and Jackson 2004; Gallese 2001), insofar as we are able to sense how others might feel about a certain situation.

As the evidence mounts for the pervasiveness of simulation as a basic process of human meaning-making and understanding, it becomes more reasonable to suppose that performing these imaginative simulations (as in Deweyan dramatic rehearsal) would give rise to emotional responses to the imagined scenarios, based on our evolved capacities to experience emotions and on the specific details of our previous emotionally charged experiences. A Deweyan account would thus build on Damasio's argument for the role of emotions in the maintenance of homeostasis (chapter 3) and also on Damasio's notion of an "as-if body loop" for running an imaginative projection, but it would take us further by providing an account of how those emotions might come into play in imaginative dramatic rehearsal. The appropriate cognitive-conative-affective simulation of possible courses of perception/action would give rise to emotional responses to the imagined situations, permitting us to assess their likelihood of resolving our problematic situation.

In short, moral deliberation is a process of cognitive-conative-affective simulation. Simulation allows us to try out, via imaginative projection, various courses of action that appear to be available in a given problematic situation, and the simulation also activates emotional responses to the projected situations. These simulations thereby permit us to give voice to various impulses, interests, and values that may be in conflict or tension and then to experience (through feeling) the extent to which a given simulation reduces, or even eliminates, the tension. What simulation accounts of meaning, conceptualization, and reasoning capture that classic representational theories do not is the fact that the simulative enactment can be the means to a tentative resolution of a problematic situation—that is, to an actual change in the situation. The "correct decision" is not available in advance of the inquiry. As we saw above, even the nature of the problem is not given in advance, but instead comes to be defined through the simulative process of inquiry itself. Nor are there always pre-existent ends given prior to the situation as a basis for means-ends calculations. Instead, we typically understand what our ends are only through our exploratory simulation of possible courses of action. Ends thus emerge and become relatively more determinate through ongoing inquiry, rather than being predefined conditions of inquiry. There are, of course, constraints on what our ends can be, but ends are not essences imposed on experience from above or beyond one's situation.

The partial account of moral deliberation I have just given recognizes a major role for feelings arising within the alternative simulations we run, as a way of exploring different possible courses of action. It is easy to anticipate the critical response of a philosopher who insists that we must explain *moral reasoning* instead of *moral feeling*. The last thing the rationalist wants to hear is that our moral deliberation hinges on imaginative projection of courses of action and the settling on one as the result of a felt transformation in the quality of our experience. How, they will ask, is this one whit different from the old '60s mantra: "If it feels good, do it"! Serial killers, pedophiles, and sadists probably all feel excited, energized, and carried forward as they commit their heinous acts. What they need, the critic will say, is absolute principled moral constraints, not merely a felt sense of the resolution of tension in their situation.

Consequently, more needs to be said about what makes a given train of moral deliberation rational or reasonable. It is to this question of deliberative rationality that I now turn.

FIVE

The Nature of "Reasonable" Moral Deliberation

The Problem of Rational Moral Judgment

The simulative process of imaginative moral deliberation that I have just described in the previous chapter treats moral reasoning as more than intuitive, non-reflective appraisal. It defines the appropriate reflective dimension as simulation. However, since emotion will play a significant role in determining what counts as a simulation that best resolves the problematic situation in a particular context, it may not be clear that there is any rationality to such processes. Therefore, I now need to explain what makes one particular choice of a simulated course of action more reasonable than some other imagined course.

Most people assume that rationality is an antecedently existing, independent, absolute structure against which we can measure our particular lines of inquiry and patterns of thinking. Indeed, many tend to take it as gospel that being "rational" means conforming to pre-given standards for right thinking, where those standards are defined by an allegedly transcendent logic and a set of absolute epistemic norms. I will argue that we ought to abandon this view of reason as a pre-given transcendent structure of thought and replace it with a more cognitively realistic account of reasonableness in moral deliberation. Reasonableness, I will claim, is an *accomplishment* of good inquiry, rather than an *a priori* set of standards governing correct reasoning.

The view of reason I am challenging is so deeply rooted

in our cultural understanding that it has for many people come to seem foundational for our Western conception of rationality. Look anywhere in the history of Western philosophy, and you will see some version of absolute reason in one major figure after another, from Plato to Aristotle to Aquinas to Descartes to Kant to Frege. According to this view, reason has an essential structure defined by universal logical principles and, sometimes, standards of epistemic justification. In the history of Western moral philosophy, what I have called "the Moral Law folk theory" assumes just such a view of a transcendent reason as the source of moral principles. It construes moral reasoning as a process in which we *discover* ultimate rational principles and *see* how they apply to particular concrete situations. Being reasonable supposedly amounts to using your innately given reason correctly. A good example of this essentialist conception of rationality is evident in Alan Donagan's roughly Kantian conception of reason:

What is assumed is that the power to reason practically may, in any rational creature, function well or ill. . . . All this presupposes that the word "reason," as it is used here, has a reference that is fixed for all possible worlds. But it seems reasonable to suppose that it has. It stands for a power, possessed by normal human adults, by which they do such things as propound propositions, assent to them or dissent from them, recognize that if a certain one is true then a certain other must also be true, and the like. What particular description is given does not matter, provided that it describes a power correctly to perform acts having contents belonging to the domain of logic; for in any possible world, one and the same power is referred to by all such descriptions. (1977, 234–35)

However self-evident this conception of transcendent reason might seem to some people, it is profoundly and dangerously misguided.[1] It mistakenly treats what is actually a conclusion and an achievement of an activity of inquiry as if it were an independently existing pre-given standard or norm. It assumes that there is some absolute structure of a transcendent logic that each suitably rational creature can utilize either correctly or incorrectly. It assumes that prior to any acts of deliberation, there already existed one correct specification of right (moral) action for a given situation, waiting to be discovered by proper reasoning.

According to the opposing conception of moral reasonableness I am proposing, an activity of moral deliberation is "reasonable" just insofar as it actually resolves the problematic situation for the better. The correctness of a deliberative process cannot be determined prior to the act of deliberation, or by any universal, pre-given standards, norms, or principles. In a typical problematic situation, we are confronted with multiple goods, values, and preferences. Our dramatic rehearsal is an imaginative simulation of how a given situation would develop were we to pursue, in turn, various available courses of action. "Choice," Dewey says, "is not the emergence of preference out of indifference. It is the emergence of a unified preference out of competing preferences" (1922, 134). As I suggested earlier, in situations of moral difficulty, we in fact have more values and preferences than we know what to do with. The problem is whether some measure of harmonization of those competing values can be effected.

What differentiates a "reasonable" from an "unreasonable" choice? An *unreasonable choice* occurs when some single value is allowed to dominate our motivational economy, without any consideration of other competing values. Then we just more or less unreflectively pursue the object or activity that appears to us to realize that value or preference. Such a choice typically proves to be unsatisfactory as a solution to a problem because attention only to one or two relevant factors of a situation is not going to allow you to harmonize the full range of competing ends and values. It thus oversimplifies the situation you encounter.

A *reasonable choice*, by contrast, is one that finds a way to harmonize to a certain degree the competing values we entertain. Dewey explains:

> But the object thought of may be one which stimulates by unifying, harmonizing, different competing tendencies. It may release an activity in which all are fulfilled, not indeed, in their original form, but in a "sublimated" fashion, that is in a way which modifies the original direction of each by reducing it to a component along with others in an action of transformed quality. Nothing is more extraordinary than the delicacy, promptness and ingenuity with which deliberation is capable of making eliminations and recombinations in projecting the course of a possible activity. To every shade of imagined circumstance there is a vibrating response; and to every complex situation a sensitiveness as to its integrity, a feeling of whether it does justice to all facts, or overrides some to the advantage of others. (1922, 135)

I regard this as the key passage in Dewey's revolutionary account of moral deliberation. In this passage we have inquiry, felt quality, and reasonableness blended and integrated. The crux of reasonable moral deliberation is the "feeling" of whether your particular situation is being composed via a rebalancing and blending of what were previously competing factors, preferences, and values. The only way to "know" that a more or less successful resolution has been achieved is to feel the harmonization of values and the opening of new, and richer, possibilities for meaningful action.

The Blending of Reason and Desire in Moral Deliberation

Moral theories grounded on faculty psychology are prone to pit various faculties against one another in a struggle for dominance. The most recurrent of these faculty contestations is that between reason and emotion, battling it out over who gets to call the moral shots. On the one side, Kant famously argued that moral imperatives can originate only from pure practical reason, with absolutely no grounding in feeling of any sort. On the other side, Hume had earlier argued the contrary view, claiming that certain refined sentiments are the source all moral action, and that reason is incapable of generating such actions. Dewey acknowledged Hume's profound influence on his moral theory (1922, 228–29), but Dewey criticized the tendency to posit the rigid separation of mental faculties, and he argued that feeling-based deliberations can be more or less reasonable. So, it is useful to see why Hume's celebrated account of moral sentiment does not tell the whole story, and to see what more is needed.

In response to the dominant exclusively rationalist moral theories of his day, Hume famously argued that moral deliberation is not really a matter of *reasoning* at all, but only a matter of *sentiment-based* judgment. Hume's argument was that understanding and reason determine "matters of fact," whereas evaluative judgments are based on emotions and the refined feelings that he called "sentiments." Hume argues:

But after every circumstance, every relation is known, the understanding has no further room to operate, nor any object on which it could employ itself. The approbation or blame which then ensues, cannot be the work of the judgement, but of the heart; and is not a speculative proposition or affirmation, but an active feeling or sentiment. In the disquisitions of the understanding, from known circumstances and relations, we infer some new and unknown. In moral decisions, all the circumstances and relations must be previously known; and the mind, from the contemplation of the whole, feels some new impression of affection or disgust, esteem or contempt, approbation or blame. (1777, 131)

Hume supplied a welcome corrective to rationalistic theories, insofar as he appreciated the central role of emotions and feelings in our moral judgments. However, because he separated reason so absolutely from emotions, he was led to erroneously conclude that there is no such thing as moral *reasoning*, if by that term we mean an instance in which reason is practical and capable of issuing directly in action. The only role for understanding or reason that Hume allowed was that of determining the facts of a situation and calculating probable outcomes of particular actions. He concludes:

> Since morals, therefore, have an influence on the actions and affections, it follows that they cannot be deriv'd from reason; and that because reason alone, as we have already prov'd, can never have any such influence. Morals excite passions, and produce or prevent actions. Reason of itself is utterly impotent in this particular. The rules of morality, therefore, are not conclusions of reason. (1739/1888, bk. III, pt. I, sec. I, 457)

Hume was right to emphasize the pivotal role of emotions and feelings in our moral deliberations, but he went too far in placing too much weight on moral sentiments, thereby denying reasoning any serious evaluative role in our moral deliberation. Having set up an absolute and unbridgeable distinction between the functions of reason and those of the emotions, Hume gave the emotions center stage in moral judgment and was thereby forced to undervalue reasoning, which he confines only to stage-setting (i.e., getting the "facts" about a situation) and causal calculation of probable outcomes of proposed courses of action. As he famously put it: "Nothing can oppose or retard the impulse of passion, but a contrary impulse.... Reason is, and ought only to be the slave of the passions, and can never pretend to any other office than to serve and obey them" (1739/1888, bk. II, pt. III, sec. III, 415).

The account of Hume that I have just summarized is the traditional interpretation, and there are commentators who argue that he has a far more nuanced and less dualistic view than many have supposed.[2] Annette Baier (1991), in particular, thinks Hume proposed a conception of *rational* sentiment, in a way that overcomes any rigid reason/emotion split. Baier's Hume is thus more compatible with the claims of Dewey (1922, 1938/1991) and Damasio (1994, 2003) that a functioning emotional system is requisite for social reasoning, and perhaps for all forms of reasoning, and that affectively based judgments can be more or less reasonable.

I am arguing that, just as emotions play a key role in our intuitive judging of right and wrong, so also our emotions and feelings are central to our more reflective moral deliberations. These deliberative processes are at once emotional, rational, and imaginative. This conception of reason comes close to the Aristotelian notion of rational desire, or desire rationally directed. It is not a question of reason versus desire, as proposed by traditional faculty psychologies, but rather a question of the reasonable ordering and harmonizing of desires. Recall the key passage from Dewey cited earlier, in which he claims that the reasonableness of a moral inquiry involves the balancing and ordering of desires:

But reasonableness is in fact a quality of an effective relationship among desires rather than a thing opposed to desire. It signifies the order, perspective, proportion which is achieved, during deliberation, out of a diversity of earlier incompatible preferences. Choice is reasonable when it induces us to act reasonably; that is, with regard to the claims of each of the competing habits and impulses. This implies, of course, the presence of some comprehensive object, one which coordinates, organizes and functions each factor of the situation which gave rise to conflict, suspense and deliberation. (1922, 135)

Reasonableness is an *achievement* of deliberative activity, something realized through inquiry and transformation of our present circumstances. We have to sense—*feel*—the order, perspective, and proportion achieved through our deliberate inquiry into the situation at hand. This feeling is akin to the way we sense or feel the achievement of a particular artwork. In a painting, for example, we may experience a blending of form, meaning, and values that enacts a world that is meaningful to us (Johnson 2007). We experience the unity and significance of the work, although we cannot properly describe this felt rightness in terms of a set of rules or by any conceptual framework. Nevertheless, we make situated judgments about the relative excellence of various artworks, although we cannot justify our claims by citing any pre-given criteria.

Similarly, in the case of moral deliberation, we also sense the working out of a pattern of inquiry in a way that transforms our current problematic situation into one that is in some measure richly reconfigured (more meaningful) and resolved (more harmonious). This observed parallel between the making (and judging) of artworks and ethical behaviors has led some philosophers to argue that we should seek to cultivate *moral artistry* in our lives by developing the virtues of creative artists and performers (Nussbaum 1986; Johnson 1993; Shusterman 2000; Fesmire 2003). As Steven Fesmire points out, "Understanding ethical reflection as artistry is supported by the nature of our wisest everyday decisions. It highlights the role of an expansive imagination that enables sensitivity to social bearings and consequences, intervenes widely and deeply in experience, and brings diverse elements together in a unified experience" (2003, 110). Moral deliberation is a process of composing life situations, and, as such, it is a form of artful remaking of experience:

The moral of the arts is that everyday moral decisions can be as richly consummated as artistic productions. The distance is narrowed between this ideal and actual deliberations to the degree that a culture focuses beyond sedimented moral criteria to education of aesthetic virtues of sensitivity, perceptiveness, discernment, creativity, expressiveness, courage, foresight, communicativeness, and experimental intelligence. (Fesmire 2003, 128)

Richard Shusterman (2000) has similarly argued that moral judgment should be modeled more on aesthetic judgment than on logical, mathematical, or scientific proof formats. Following Bernard Williams (1985), Shusterman stresses that ethics cannot be narrowly confined to a pregiven system of moral obligations. Once we realize that ethics concerns a far broader and more variegated body of goods and values relevant to a well-lived life, we come to see ethical deliberation as more like making and evaluating artworks than the application of universally binding moral principles:

> Demoting moral obligation to merely one significant factor in ethical deliberation on how to live the good life makes such deliberation much more like aesthetic judgment and justification than syllogistic or legalistic reasoning. Finding what is right becomes a matter of finding the most fitting and appealing gestalt, of perceiving the most attractive and harmonious constellation of various and variously weighted features in a given situation or life. . . . Likewise, ethical justification comes to resemble aesthetic explanation in appealing not to syllogism or algorithm but to perceptually persuasive argument (through well-wrought narrative, tendentious rhetoric, and imaginative examples) in an attempt to convince. . . . For ethical decisions, like artistic ones, should not be the outcome of strict application of rules but the product of creative and critical imagination. Ethics and aesthetics become one in this meaningful and sensible sense; and the project of an ethical life becomes an exercise in living aesthetically. (Shusterman 2000, 244)

I will return to the nature and role of ethical argument and justification below. My focus now is on the way moral deliberation is an activity of transformative thinking that reconfigures the situation by ordering the materials of that situation into a new gestalt. Our sense that our imaginative reflection is right (or better, or best) *at this present moment* comes through the transfiguration of life, that is, from the felt resolution of our difficulties. Reasonableness (or reasonable desire) is thus a consummatory achievement, however temporary or tentative it might be. It requires the laborious cultivation of sensitivity to the many conflicting factors and values within a situation. It requires an apprenticeship in reading circumstances and composing harmonious situations. It requires the educating of desire and the sensitizing of reason. That is why Dewey defines reasonableness as an achievement, rather than a given: "'Reason' is not an antecedent force which serves as a panacea. It is a laborious achievement of habit needing to be continually worked over. A balanced arrangement of propulsive activities manifested in deliberation—namely, reason—depends upon a sensitive and proportionate emotional sensitiveness" (1922, 137).

The Critical Dimension of Moral Deliberation

What prevents imaginative moral simulation from devolving into a free-for-all contest of competing urges and values? Why does it not collapse into "if it feels good, do it"? In other words, by virtue of what can there be any *critical* dimension to our dramatic rehearsal? We need an account of excellence in inquiry. The antidote to exclusive attention to one particular perspective, emotion, principle, or value is to open oneself to the range of perspectives, emotions, principles, and values at work in any given situation. This is what is meant by moral perceptiveness, sensitivity, and openness. It is a cultivated capacity to "see what is going on" and to discern the relevant considerations. A major challenge in moral deliberation is not to allow any single factor exclusive sway in our reflections, without first considering a broad range of other relevant present factors. For example, a natural response to some wound or harm, whether physical or emotional, might be a passion for revenge. However, one can follow out a path of revenge via moral deliberation and see where that path might lead. Other considerations—such as justice, kindness, concern for the greater good, desire to help others, importance of cooperation, and so on—provide a counterbalance to the hegemony of a single factor. Here is how Dewey so elegantly frames the issue:

Rationality, once more, is not a force to evoke against impulse and habit. It is the attainment of a working harmony among diverse desires. "Reason" as a noun signifies the happy cooperation of a multitude of dispositions, such as sympathy, curiosity, exploration, experimentation, frankness, pursuit—to follow things through—circumspection, to look about at the context, etc., etc. . . . Reason, the rational attitude, is the resulting disposition, not a ready-made antecedent which can be invoked at will and set into movement. The man who would intelligently cultivate intelligence will widen, not narrow, his life of strong impulses while aiming at their happy coincidence in operation. (1922, 136–37)

Why, one might ask, ought a moral inquirer to *widen* her perspective? Why should breadth and richness of perspective trump other values? And would that not lead us back into just one more fundamentalism—a fundamentalism of a single value—namely, pluralism of perspective?

The answer, I submit, is that pluralism is the method of including in your inquiry and deliberation the broadest possible range of competing values and concerns, and so it is not a single value that might contest with other values. Pluralism is the view that since there are no absolute, non-perspectival values and standards, our best strategy of inquiry is to expand our range of perspectives and to seek the richest description of the complexities of the situation that we can muster. It is to seek what Steven Winter calls "transperspectivity." Absent any human access to a God's-eye view, "'impartiality,' in turn, is no longer a matter of an aperspectival position, but rather an exercise of the empathetic ability to imagine what a question looks like from more than one side" (1990, 685). Since there is no single all-encompassing framework or all-devouring method, we ought to search for converging evidence that arises from multiple perspectives and methods of inquiry.

The only way to constrain our prejudices is to entertain other prejudices—other prejudgments emerging from other points of view. Dewey is often falsely accused of scientism, because of his repeated embrace of the methods of empirical scientific inquiry as the key to intelligent practice. But Dewey was not reducing all modes of knowing to fields of science; rather, he was holding up experimental methods of hypothesis and testing as models of intelligent inquiry for creatures who are not omniscient, but always perspectival and fallible. The only way to overcome the limits of our parochial perspectives is to bring other perspectives into play and hopefully to seek commonalities shared by different orientations and viewpoints.

Paul Feyerabend (1975) was fond of urging us to let a thousand flowers bloom, as the best strategy of inquiry in a post-Kuhnian world where every perspective is encompassed by some particular paradigm formation. True to his convictions, Feyerabend would throw medieval alchemy into conversation with modern chemistry, and early Greek cosmology into dialogue with contemporary quantum physics. His point was not that he would stake his life on Greek medical views about the balancing of animal spirits (black bile, yellow bile, blood, phlegm, etc.) in preference to modern medical practices, but only that the consideration of multiple perspectives is the only way to recognize the framing assumptions and limitations of our own preferred, and often uncritically accepted, views on any subject. For example, medical orthodoxy has recently become more receptive to homeopathic (holistic) practices. This, in turn, allows us to recognize certain limitations of Western allopathic (cut, burn, poison) medicine. Feyerabend's rage "against method" (which was the title of his most famous book), is not against the utilization of multiple methodological frameworks, points of view, and modes of inquiry, but rather a rejection of any hegemonic conception of *the* method of knowing.

Feyerabend directed his analysis and criticism mostly at scientific and philosophical theories, but his arguments apply equally to claims to moral knowledge. Once you understand the limitations of human cognition, a pluralism of moral values and perspectives is inescapable. However, that does not mean that "anything goes," any more than it means that every scientific theory or perspective is equally insightful.

Moral deliberation is a series of simulations conducted under the influence of varied values, ends-in-view, principles, and other factors (such as notions of the virtues and conceptions of life well lived). Moral principles should be understood as summaries of considerations that some community of inquirers has found important to factor into their previous moral deliberations. Each principle typically represents a way of privileging a single value over other possible values that might be operative in the situation. A principle such as "Thou shalt not kill," taken literally, treats all life (or, according to some interpretations of this commandment, all human life) as an absolute value, one capable of overriding any other values that might possibly conflict with this supreme value (e.g., success, honor, military duty, revenge, justice, or even obedience to various religious imperatives, such as "He that blasphemeth the name of the Lord, he shall surely be put to death, and all the congregation shall certainly stone him" [Leviticus 24:13–16]). Moral principles, in short, are condensations of fundamental values of an individual or group.

My conception of the critical dimension of moral reasoning can now be stated as follows: To assess a projected simulation (where each simulated course of action manifests at least one possible principle or value) is to run the simulation as a dramatic rehearsal, with the values inherent in the proposed principle being given more weight in the structuring of the developing situation than other values. In other words, the simulation reveals how life might develop if that principle or value is given priority. Each simulation we run will typically evoke in us a felt sense of how certain values or standards would more or less successfully unify the situation as we currently understand it. Other dramatic simulations will reveal the meaning and implications of other principles and values. Each simulation is a different enactment of the possibilities for developing and reconfiguring our present conflicted experience.

To state the obvious, one thing that makes genuine moral problems so difficult is precisely that human existence is a complicated affair, so that all of our attempts at simplification are doomed to miss much of what should matter in deciding how to act. If a situation were easy and straightforward, we would not need to deliberate, because our normal habits of valuing would carry us forward without the need for any forethought. Simplification is but one tool in our toolbox of strategies for inquiry and problem-solving, but it is not the only tool, and it ought not to be used when a complex situation calls for other tools. One of the greatest sins of fundamentalism is the exclusive utilization of simplification and reductionism as the sole strategy for inquiry. The charge that fundamentalists are unthinking is justified insofar as their quest for simplification and absolute certainty blinds them to other perspectives, interests, and values operative in complex situations.

The account of good moral deliberation I have just given could provide one possible construal of what Rawls (1971) meant by "wide reflective equilibrium." *Narrow reflective equilibrium* requires us to test our moral intuitions, our view of human nature, our favored principles, and our preferred institutions against the broader cultural system in which we find ourselves. This includes taking a more comprehensive historical view of the development of our shared communal sense of justice and morality, understanding the particular historical conditions that gave rise to it, and determining whether current conditions are significantly different so as to require a new normative assessment. *Wide reflective equilibrium* moves us beyond our own cultural tradition to a consideration of other moral traditions that might call into question some of our most taken-for-granted assumptions (Flanagan 2007, 121–25). For example, narrow reflective equilibrium would be sufficient for a devotee of the Aryan Brotherhood to realize that his racist doctrines are completely incompatible with contemporary democratic notions of racial equality, tolerance, and justice. Wide reflective equilibrium would provide us with a broader global perspective that would make it possible for us to criticize aspects of our more parochial beliefs and practices that we have mistakenly assumed to be universal. For example, wide reflective equilibrium would reveal internal tensions and conflicts sustained by our particular American version of democracy-*cum*-capitalism that is built upon the maintenance of inequalities and injustices perpetrated under the guise of certain conservative conceptions of democratic freedom. This would allow us to see how one particular American conception of freedom is highly problematic when considered in a broader global perspective.

There can be no overarching principle for deciding when a sufficiently wide perspective has been achieved. It is often the case that criticism of a long-standing moral institution or system is made possible precisely when someone or some group subsequently delineates a larger, more inclusive, boundary around "who counts" in a moral/legal/political/economic determination. Pre-Civil War slave owners were mostly at peace with their institution of slavery. Slaves were not considered to be fully human, and so they could be appropriated as animals for the satisfaction of their owners' purposes. God was in his heaven, and all was right with *their* (the slave owners') world, although obviously not with their slaves' world. There was a broader, more inclusive perspective that would have called into question some of the founding assumptions of the institution of slavery, but that wider reflective point of view was more or less successfully repressed from the consciousness of those in power. A painful history of suffering and national conflict has now led us toward a broader, more comprehensive point of view that includes African Americans as full human beings. Although some of our institutions, practices, attitudes, and laws may be in conflict with this vision, there is at least a slow drift toward recognition of some level of human equality. We obviously have not achieved this full awareness of racial equality, but my point is that normal processes of narrow and wide reflective equilibrium are making it possible to move us in this direction. The question of who is accorded moral personhood—that is, who counts as a moral being—is still a work in progress that is part of our changing moral landscape. Moreover, there are now some who argue for the attribution of moral personhood and dignity, not just to humans, but to some sentient non-human animals and even to plants, ecosystems, and the cosmos conceived as one massive ongoing creative process.

There can be no absolute, fixed, and final answer to questions of this sort, because that would presuppose (1) the positing of a finished, complete, unchanging physical, interpersonal, and cultural reality, and (2) an aperspectival, all-inclusive grasp of all that is. Both of these assumptions are equally false and misleading. Reality is not a completed whole, and we have no God's-eye view of its essential nature. We are ineliminably perspectival creatures who find themselves embedded in processes of a changing world. We are where we find ourselves in our current debates, wars, cooperations, and contestations. We are immersed in processes of ongoing moral conflict, inquiry, and argument from which we cannot hope to extricate ourselves. These debates include not just questions about our most fundamental values or principles, but also questions about the proper conception of moral inquiry, and what a moral theory ought to do for us.

We need to get over any anxiety that abandoning our deepest absolutist aspirations would entail losing something crucial for moral judgment. As we will see in chapter 7, we *never actually had* strict rules or absolute principles for deciding moral problems once and for all. We *never had* absolute moral criteria given prior to our deliberation and judgment. We *never had* an autonomous, radically free will of the sort required by the Moral Law folk theory. It is necessary to learn to live without something we never really had in the first place and to stop longing for something that is not possible from a human perspective. We need to embrace a morality fit for humans.

It has often been noted that Galileo (following Copernicus) turned everything upside down by removing the earth from the center of the universe, thereby symbolically removing *us* from the center of God's creation. Next, Newton jolted us by arguing that natural events are explainable by efficient causes, without the need to posit final causes (ends or purposes established either by God or nature). This left many feeling abandoned and alone in a supposedly purposeless, meaningless universe of causally determined objects and events. Darwin then delivered another blow to our self-image by bringing us down to the level of (highly intelligent) animals, and Freud later challenged our rational superiority by showing how much of our thinking is driven by unconscious drives and primary processes. Today our existential anxiety about our progressively diminishing status in the universe has been intensified among those who view recent discoveries in the cognitive sciences as threatening to dismantle our moral foundations. They perceive, correctly, that some cognitive science research undermines at least one of our last great holdout ideas, namely, the idea that moral reasoning is a transcendent, rational, deliberate, conscious, and free process of autonomous thinking. It is hard to give up the ideal of a transcendent moral order against which we can measure our intentions and actions, but that is what we must do, in light of our dawning awareness of the nonconscious nature of most of our moral judgment, the changing character of experience, the vast role of feelings in moral cognition, and the non-formalizability of imaginative moral deliberation.

However, even though the cognitive sciences have dethroned one image of humanity (i.e., as little gods with transcendent rationality and a spark of the eternal), they have replaced that idol with an image of humans as moral inquirers possessed of cognitive, affective, and imaginative resources at least somewhat adequate to the types of problems we encounter in life. We thus need to show how it is possible to think of progress in moral growth.

Kitcher's "Ethical Project"

From a perspective quite similar to the one I am articulating here, Philip Kitcher, in his book *The Ethical Project* (2011), has developed a pragmatic naturalist account of the nature and continuing evolution of human moral systems. Kitcher observes that human systems of morality emerged originally in small groups attempting to find ways to live together cooperatively and to flourish as individuals and communities. Part I of his book is an extremely insightful and well-argued set of speculations about how it might have been possible for our ancient ancestors to develop forms of altruistic group behavior that would have supported the survival and well-being of early societies.

One of Kitcher's central unifying themes is that our forebears *began* an "ethical project" that has continued down to the present day, and that will continue into the future, so long as there is human life. It is a project of ongoing moral inquiry, subject to revision as newly emerging conditions demand. A key part of that project is the determination and reassessment of the types of functions that moral systems emerge to satisfy, and how well various systems accomplish those tasks. Kitcher nicely summarizes his key claims about these functions in the following passage:

Pragmatic naturalism makes the following claims about the original function of ethics. (1) The problem background consists in social instability and conflict caused by altruism failures. (2) The original function of ethics is to promote social harmony through the remedying of altruism failures. (3) Our ethically pioneering ancestors had only a dim appreciation of the problem background, responding to the difficulties and discomforts of a tense and fragile social life. (4) We know more about the problem background than they did and can offer partial and incomplete diagnoses of the types of altruism failures to be remedied; we can understand the success of certain kinds of ethical systems in terms of these partial characterizations. (5) Even with respect to the original function, the project of refining the codes we have continues. (225)

Kitcher is proposing an ongoing process of moral inquiry of the sort I have been developing here. Societies engage in moral dialogue over what types of functions moral systems are meant to address, how various problems are to be understood, what counts as an acceptable answer or source of moral guidance, and what criteria, if any, there are for moral progress. Notice that even what counts as a moral function or a moral problem is subject to scrutiny and reconsideration. Nothing is set in stone, but some dimensions of our moral systems will be so stable as to constitute virtual givens. I am thinking here of very basic assumptions, such as that one function of all moral systems must be to coordinate group interactions and to harmonize, as far as possible, conflicting values. Our conceptions of how best to carry out this function will vary over time, but the human need to satisfy this particular function seems built into the very nature of our moral situation.

The "dialogue" between competing views involved here is not merely a matter of words spoken between interlocutors, for it will also include actions of individuals and groups that might challenge the established practices and institutions of a society. In other words, the "ethical project" is not just a matter of incessant talking, debating, and arguing. It is an *experimental process* of trying out various modes of behavior (verbal and nonverbal), various institutional structures, and various life strategies. It consists of on-the-ground experiments in living. For example, contemporary practices explored in home schooling can reasonably be viewed as experiments in ways of living. As such, they are forms of embodied, embedded, enactive inquiry that constitute "arguments" for and against certain practices as ways of addressing the kinds of problems we have identified for ourselves.

How, then, do we decide which of several experiments in living is to be preferred at a given point in history? Eschewing notions of ethical truth, Kitcher argues that we should instead focus on the conditions that make possible moral progress. For any two conflicting codes of conduct *"to salvage the notion of 'objectively better than' that occurs in these claims and counterclaims, we do not need any concept of ethical truth. It is enough to recognize which kinds of changes would be progressive or regressive"* (210; italics in original). Much of the last half of his book is devoted to setting out some of the considerations relevant to an assessment of which code is more progressive.

One of Kitcher's key criteria for progress is to always seek to widen what Peter Singer (1981) called "the expanding circle" of beings affected by individual or group actions and practices. In my brief earlier comment on changing views of chattel slavery, it was precisely this procedure of recognizing a wider scope for the term "moral person" (or even "human being") that proved to be so essential in bringing about the abolition of slavery. Singer famously (or infamously—take your pick) argues that we ought to expand the boundary of moral personhood far beyond its present confines to embrace, at the very least, many non-human species. There can be no algorithm for determining when the circle of relevant interlocutors and affected parties has been drawn broadly enough, because where that line is drawn is itself the subject of ongoing evaluative/normative debate. Typically, it is only after the fact, judged from the perspective of how things now look because we earlier expanded the circle, that we can see the wisdom (or folly) of having included new, previously unrecognized or marginalized voices.

Another important criterion of progress is the ability of one system to recognize and deal with an expanded or altered set of desires and values that have come into play because of changing material and cultural conditions. The ever-increasing complexity of life gives rise to new possibilities and new desires (and, hence, new problems) that could not have been imagined by some of our ancient ancestors. New technologies generate new possibilities for human behavior plus new desires and values. In this context, that code is more progressive that can understand how the new desires/possibilities/values arose, why they were not part of the deliberations of earlier groups, and how we now can accommodate these new desires or values within our evolving moral system. For example, new technologies can produce new kinds of objects, which, in turn, can generate hitherto unanticipated values and desires.

For my purposes, the upshot of these and other of Kitcher's desiderata for assessing moral progress is that "the ethical project evolves indefinitely. Progress is made not by discovering something independent of us and our societies, but by fulfilling the functions of ethics as they have so far emerged. The project is something people work out with one another. There are no experts here" (285–86). As Alasdair MacIntyre argues in the postscript to the second edition of *After Virtue* (1984), one key dimension for determining the superiority of one tradition over another is that the tradition judged to be superior (i.e., more progressive) can explain both (1) why the "inferior" tradition could neither fully appreciate nor solve certain problems, and (2) how the replacement tradition can either dissolve or resolve some of the tensions inherent in the prior tradition. However, given the limitations of human resources, intelligence, and creativity, there is no assumption that a moral system judged to be superior will solve all of the problems left by its predecessor tradition, nor that it will not itself generate some of its own distinctive problems that it cannot solve. Indeed, it is for that reason that the ethical project is an unending task.

On those occasions where we rise to the level of theoretical reflection on competing moral alternatives, there will be certain types of necessary idealizations that we must use to assess progress. For example, we will sometimes need to formulate a utopian vision of a moral system or institution, in which everyone is treated equally with regard to available resources, offices, chances to pursue their individual interests, and so on: "In this state, each member of the human population has a serious chance of living a good life, a life in which the person can recognize a number of different possibilities for living, can make a free choice of a project informed by that recognition, and realize a significant number of the plans, intentions, and desires across the population" (Kitcher 2011, 316). Of course, this is a dramatic idealization that is probably nowhere to be met in actual human societies. However, we can often usefully employ such ideals as a way to assess the degree to which different moral systems come close to or diverge from various aspects of the articulated ideal.

Another very important consideration regarding the assessment of moral progress is the question of who our interlocutors are in the discussion of better and worse systems or codes. In earlier small-scale societies, it might have been possible (though, I suspect, exceedingly rare) that a group of relative equals could engage one another regarding the nature of the problems faced, the types of functions served by morality, and the best way to satisfy those functional needs. Very early in human history, however, this would have become completely unworkable, due to the rapidly increasing size of the community. Consequently, today we hardly have any mutual engagement of this sort, even with people in our apartment building, on our block, or in our neighborhood, not to mention the impossibility of genuine democratic dialogue at the city, state, national, or global level. Necessarily, then, we must idealize our interlocutor, simulating an ethical engagement as if we could address a representative group of our fellow citizens of the world. We are forced to simulate a conversation in which we introduce "considerations and lines of reasoning that would be brought forward to achieve consensus were the entire human population to participate, *under conditions of mutual engagement,* in a conversation about the regulation of conduct" (Kitcher 2011, 340).

Anyone who is even the least bit familiar with twentieth-century political theory will, of course, recognize a Rawlsian perspective evident in Kitcher's view about the nature of argument and justification. Like nearly all non-foundationalist social theorists, Kitcher appropriates some version of John Rawls's famous notion of "reflective equilibrium" as a process for constructing a theory that is sensitive and responsive to actual lived conditions, avoids any claim to absolute grounding of values or principles, and gives us a means of proposing moral, social, and political institutions that fit the circumstances of our present situation. One need not accept (as I do not) all of the particular assumptions that Rawls makes about the initial conditions within which various possible social, political, and moral practices and institutions are proposed and evaluated; but one can approve of Rawls's general conception of the decision and justification procedure in which no part or assumption is taken as absolutely unrevisable or chiseled in metaphysical stone. Good moral reflection requires us to remain open to the possible questioning of *any* aspect of our received view. That will include views about what the basic ethical functions are, how we circumscribe the relevant phenomena we must consider, what the most likely outcomes are of various acts and practices, what constitutes a good life, and so on.

In short, Kitcher's claim that humanity has embarked on an ongoing ethical project from which we can never escape, so long as humans draw breath on this earth, is simply a consequence of our status as evolving, changing creatures, who often encounter novel emergent conditions hitherto unanticipated. The fact that we can never pretend to have knowledge of absolute truth in no way keeps us from making reasonable claims about the superiority of one set of values, principles, and practices over another.

The Descriptive-Genealogical-Nomological Component of Moral Systems

Approaching ethical deliberation as a complex form of ongoing *problem-solving* removes morality from the traditional framework that looks for an ultimate grounding in some allegedly universal set of categorical principles, instincts, virtues, or moral feelings. The alternative, naturalistic approach consists of an empirically and experimentally grounded inquiry into forms of human well-being, virtues, and ethical principles. We need an empirically informed approach that allows us to understand our values and principles, criticize them when necessary, and imagine alternative perspectives as a way of solving our pressing moral problems. Such an approach would locate the source of what we call "moral" values, principles, deliberation, and judgment in processes of organism-environment interactions, rather than in God, pure reason, or a unique moral faculty. Eschewing claims to "pure" concepts, "*a priori*" forms of knowledge, or totalizing methods, naturalistic approaches draw on all sources of knowledge currently available. Those include biology, anthropology, neuroscience, and psychology, but also philosophy, history, literature, and the arts, in order to explain how people come to value certain activities and states, to explore how cultures develop their distinctive moral systems, and to imagine new forms of moral engagement with one another.

CHAPTER FIVE

Naturalized views need not deny the possibility of cross-cultural ethical principles (or, at least, ethical dispositions), but they see no need to try to ground them in alleged universal human faculties or instincts.

Naturalized orientations thus treat the study of morality as what Dewey called a subject of empirical and experiential inquiry:

> But in fact morals is the most humane of all subjects. It is that which is closest to human nature; it is ineradicably empirical, not theological nor metaphysical nor mathematical. Since it directly concerns human nature, everything that can be known of the human mind and body in physiology, medicine, anthropology, and psychology is pertinent to moral inquiry. (1922, 204)

Today one might be tempted to add cognitive neuroscience to Dewey's list of empirical disciplines, but that would not change his point, which was that we ought to think about morality concretely and empirically. Instead of trying to "ground" our moral ideals in some allegedly universal or absolute source, we ought to ask how it is that creatures like us, inhabiting the kinds of physical, interpersonal, and cultural environments we do, come to have the values we do, and how it is possible for us to put them to the test of our ongoing personal and communal experience.

Owen Flanagan has suggested that a suitably naturalized ethics will have two basic components, both of which will benefit from extensive empirical inquiry. The first is a *descriptive-genealogical-nomological* component, and the second is *normative*. The descriptive-genealogical-nomological component is based on an empirical study of basic propensities of *Homo sapiens* and on the history of our moral behavior and institutions. The point of this is to understand the sources and development of our moral values and practices (and those of other cultures). It will explain how various peoples tend to circumscribe the domain of ethics (Shweder and Bourne 1984; Shweder, Mahapatra, and Miller 1987; Haidt 2012), but it will not regard such circumscriptions as rigid, metaphysically given, unrevisable, or impermeable. It will ask why beings like us might develop the kinds of values we have and how we are prone to understand those values. My survey of sources of moral values (chapter 2), my account of intuitive processes of moral judgment (chapter 3), and my theory of moral deliberation (chapters 4 and 5) represent a significant part of my contribution to this descriptive-genealogical-nomological component. As we have just seen, parts I and II of Kitcher's book (2011) give his account of these descriptive-genealogical-normative dimensions of our ongoing human ethical project. Part III constitutes his account of the normative.

THE NATURE OF "REASONABLE" MORAL DELIBERATION

Robert Hinde's *Why Good Is Good: The Sources of Morality* (2002) is an excellent specimen of what a naturalist's descriptive-genealogical-nomological component might involve. It presents an eminently reasonable picture of a general human nature built on certain widely shared bodily capacities, emotional needs, psychological characteristics, and mental operations that give rise to very general moral principles. Hinde identifies several human potentials and their correlative psychological characteristics, and he argues that these tend to give rise to some very general moral principles, which, in turn, support the more culturally specific (and therefore more variable and non-universal) moral precepts that flesh out those abstract ideals:

All humans are born with certain *potentials*, which, in interaction with the experienced environment, give rise to psychological *characteristics*. . . . The psychological characteristics include propensities to behave in some ways rather than others, and predispositions to learn some things more readily than others. An example of such propensities might be the tendency to behave in a positive, rather than selfishly assertive, way to friends, and an example of a predisposition might be the ability to learn the language generally spoken in society. . . .

Some apparently basic moral themes that arise in (virtually) all cultures are referred to as *moral principles*. That parents should look after their children is an example. . . .

Moral *precepts* (and values) are more specific guides to behaviour, which may be either explicit (as in the Judaeo-Christian Ten Commandments) or implicit, as in the behaviour of individuals. (12–13)

Notice the significant difference in tone between the assertions of those who believe there are transcendent sources of values and absolute moral principles and the more cautious and empirically sound descriptions offered by Hinde. The former talk of discrete kinds of experiences, types of judgments, and, as we will see in the next chapter, a moral faculty and a universal moral grammar. Hinde talks of propensities, tendencies, characteristics, and "virtually" universal general principles. The difference in tone may seem ever so slight, but it is philosophically profound. It marks the primary difference between absolutist claims of transcendent sources of values or principles, on the one hand, and the ethical naturalist's observation of tendencies and general propensities, on the other.

Ethical naturalists recognize constraints on the forms a morality can take, based primarily (though not exclusively) on what the cognitive sciences are revealing about human needs, cognitive capacities, emotions, values, judgments, and social interactions. Some of these constraints arise from the way our brains and bodies work, others stem from the requirements for interpersonal nurturance and cooperation, others are based on accounts of well-being, and still others are shaped by the nature of our various communal/cultural relations and practices. Hinde nicely summarizes this multidimensional approach:

> Both culture and morality stem ultimately from 'human nature' as it has been shaped by natural and cultural selection in interaction with the physical, biological and social environments that humans have experienced in evolutionary and historical time and that are experienced in the lifetime of each individual. The important word here is 'ultimately.' There is no implication that moral precepts are simply there in our biological heritage or in our experience: both are essential. Thus it will be argued that moral codes are constructed, maintained, transmitted and amended by humans interacting with each other, and thus depend on human nature (in the restricted sense discussed below) and on experience in the physical, psychological and cultural environments of development. If this thesis is accepted, it renders unnecessary any appeal to a transcendental source. (2002, 13; italics in original)

Hinde's "human nature" is not a fixed essence, but rather an evolved and evolving set of characteristic needs, capacities, and propensities that give rise to what we call moral concerns. A psychologically realistic moral theory that is naturalistic will have to overcome the temptation to regress back to the postulation of transcendent moral sources of values, in the guise of alleged moral faculties, organs, and instincts. Instead, it will base its descriptive-genealogical-nomological account of morality on the best empirical research accessible from all available disciplines.

Better and Worse: The Normative Component of Moral Systems

The second part of a naturalized moral theory is the *normative component*, which will "explain why some norms—including those governing choosing norms—values, virtues, are good or better than others.... The normative component may try to systematize at some abstract level the ways of feeling, living, and being that we, as moral creatures, should aspire to. But whether such systematizing is a good idea will itself be evaluated pragmatically" (Flanagan 1996, 21).

Flanagan's characterization of morality is similar to Dewey's in being de-theologized, de-transcendentalized, experiential, and, above all, experimental:

> Morals consist of habits of heart, mind, and behavior. Morality is "normative" in the following sense: It consists of the extraction of "good" or "excellent" practices from common practices. Ethics consists of wisdom based on historical experience about how best to arrange our affairs, and how to develop our nobler potential, as this too is judged on the basis of historical experience. Moral habits, wisdom, and skills consist largely of "know-how" that allows for smooth interpersonal relations, as well as for personal growth and fulfillment. Ethical reasoning is a variety of practical reasoning designed to help us negotiate practical life, both intrapersonal and interpersonal, as it occurs in the ecological niche we occupy. (2007, 126)

Experimental and experiential approaches to morality that reject claims to absolute foundations for moral capacities and values promise accounts that are psychologically sound, historically situated, and sociologically informed. However, the non-absolutist, context-dependent character of naturalized ethics remains for some people a source of uneasiness, insofar as they believe that without universal principles or values we are thrown into moral, social, and political chaos. Having given up supernatural foundations, and also foundations in any form of transcendent or universal reason, they still seek (as we will see in chapter 6) at least some foundation in biological (brain-based) capacities, to which they give the name *moral faculty*, *moral organ*, or *moral grammar*.

I believe that the motivation for these attempts to find ultimate moral foundations is a profound anxiety over an imagined rampant moral relativism. The pragmatic naturalist must address the issue of normativity head-on. How can we get any normative leverage, if all we, as naturalists, seem capable of offering is empirically responsible descriptions of the nature and sources of moral systems? Where does the normative enter into the picture?

Owen Flanagan and Robert Williams (2010) have provided the best answer to this perennial question that I have encountered. Their argument, in brief, is this: Values (and "oughts") are never *deduced* from factual statements. Instead, our normative assessments are based on abductive inferences about which values would best resolve our current moral problems. They become hypotheses that we try out in our imaginative simulation of a projected course of action. What we *ought* to do is a question of what is reasonable, and reasonableness can only be determined within a community of inquirers and practitioners, given standards recognized by that community. We can, on occasion, call into question some of those standards, but never from a God's-eye absolute perspective. Here, in more detail, is their argument:

CHAPTER FIVE

1. Hume was right in observing that from no set of descriptive factual statements alone can you deduce a precept about what ought to be done. For example, "That you ought to eat breakfast does not follow from any set of facts about you and nutrition. Is it a good idea to eat breakfast? Of course. Can you derive the conclusion that you ought to eat a healthy breakfast from facts about nutrition, plus facts about your own desires for health and well-being, plus those of your loved ones? No. Minimally, you will need to add a premise to the effect that one ought to do what achieves one's ends. And then the problem repeats. How does one justify this 'ought'?" (Flanagan and Williams 2010, 442–43).
2. The most important things in our lives tend not to be susceptible of proof or deductive demonstration, anyway, and so it is high time that we get over the illusion that we need proof or rational deduction to justify the "oughts" in our lives.
3. Most of our inferences about what ought to be done are either inductive (i.e., based on past experience of what has worked well) or abductive (i.e., involving inference to the best explanation among the alternatives that occur to us). Moral decision-making is a form of problem-solving, and Flanagan and Williams observe that accomplished communities of problem-solvers (e.g., city planners, bridge builders, architects) discuss among themselves the problems they face. Then they try to come up with good ideas for possible solutions, and they design ways to accomplish these ends-in-view they have formulated. They do not need any transcendental justifications to ground normative claims of this sort.
4. Any "oughts" that emerge from such problem-solving discussions arise within the context of the practices in which the participants are engaging. And there is no miracle in that, since these people have spent their lives trying to devise "best practices" relative to the limitations of their resources, knowledge, and opportunities for action. The crucial point here is that "'oughts' are not (normally) derived or deduced. But 'oughts' do not sit out in thin air in such a way that only the ghost-whisperers can explain. 'Oughts' are reasoned to in a holistic network that operates over both propositions about facts and propositions about antecedently settled values and 'oughts,' all of which are open to conversational challenges. Neither the ends nor the means are deduced" (444).
5. Consequently, discussions about ends and "oughts" are matters of what is reasonable, and there can be no absolute, *a priori* standards of reasonableness. Reasonableness, like justification, depends on who is talking to whom, and under what conditions. Reasonableness depends on finding shared assumptions and values. Reasonableness is a matter of how a certain view or practice enriches, harmonizes, and liberates our action in the world. Dewey argued that ethics is like farming, in which better and worse practices emerge as a result of long-term practice and experimentation. Flanagan and Williams argue, in a parallel fashion, that ethics is like engineering: "Specify ends. Talk about whether ends involve good, worthy goals.

Specify means, evaluate them in terms of effectiveness, costs, and so on. Then do the right thing. Everything, means and ends, is open to conversational challenge" (444).

Especially when it comes to moral matters, most of us would like to believe that there are truly knock-down, drag-out arguments for moral values, principles, and virtues. But we can never get around the glaringly obvious fact that what constitutes a "knock-down, drag-out" will depend on the community of inquirers (or moral agents) with whom you are engaging. Though I hesitate to give G. E. Moore credit for any insight about morality, he was at least correct to observe that for any statement of the form "X is good," it is always possible to turn around and ask, "Why is X good?" In other words, we cannot really argue our way to moral conclusions by explaining why something is good. Moore thought that this "open-question" argument was somehow devastating for moral theory, because he believed this made it impossible to define good. He concluded, quite mistakenly, that good was an indefinable nonnatural property. Fortunately, demonstration, proof, and rock-bottom definition, as Flanagan and Williams observe, are not what we are up to in discussions of morality. *All we have, and all we have ever had, is the best argument we can muster, within the community of our interlocutors, for the reasonableness of our values and proposed ends (and the means for realizing them).* Some of these "arguments" are not stated in words, but are rather revealed in the experiential outcomes of various experiments in social arrangements, approved practices, and visions of human flourishing.

Flanagan makes a similar argument about normativity when he says that we must accept our "internalist" predicament of never having access to any unconditional, absolute, or value-neutral conception of any moral notion. Nevertheless, we can assess the relative merit of competing conceptions of well-being and moral improvement, although *never* absolutely, or for all time:

We are always, because it is the nature of the human condition, in some kind of internalist predicament. But normative assessment and adjustment need not arise internally in a narrow objectionable sense. Normative assessment, moral improvement, and so on, must come from inside the dialectical space of meta-norms for resolving disagreements or deciding to live tolerantly without agreement. This means that the internalist predicament admits of degrees and that to the degree we reflect widely, from a perspective that includes other moral conceptions we are going external in some meaningful sense of the term 'external.' (2007, 138)

Narrow and wide reflective equilibrium, like reasonableness, are not notions that can be defined by a rigid set of absolute standards. There can be no context-free determination of when one has achieved a sufficiently capacious and broad perspective. Nor can there be any algorithm for deciding what is problematic in our received view and ought to be changed. Those who think otherwise have not paid attention to what we have been learning about the conditions for human meaning, understanding, thought, and reasoning (see chapter 7). Consequently, wide vs. narrow equilibrium must always be a matter of degree, and reasonableness is a contextually determined notion subject to the practices and values of particular communities of inquirers.

The key, then, to understanding moral progress is to recognize that what we call "moral" practices are really on a par with other forms of complex human practice, such as farming, medicine, engineering, and art. In all of these forms of practice, we are faced with problems, both old and new, that call for response. We have developed moral systems and practices for addressing certain problems, but new problems arise that call for a critical perspective and creative inquiry. In all forms of practice, there are ends, means, values, and institutions. Although these may be given or assumed in the context of our current practices, the point is that neither our ends, means, nor values are in principle fixed or unrevisable, even if there are plenty of good candidates for universal or near-universal values, as I argued in chapter 2. Problem-solving is not, therefore, merely means-to-ends reasoning. Instead, the very ends themselves, and the values they give evidence of, have to be subject to scrutiny and possible reconstruction. In this way, our most basic notions of a "good life" are the focal subject of our moral investigation. We can assess these notions, not because we have some tap into eternal truths, but simply because we have capacities for analysis, inquiry, and creative problem-solving.

SIX

There Is No Moral Faculty[1]

I have been arguing that the past thirty years have witnessed a dramatic and important change in the landscape of moral theory. The principal difference between then and now has been the emergence of a vast and rapidly growing body of naturalized approaches to the origins and nature of moral judgment. As I reflect back on my education as a philosophy graduate student in the 1970s, I recall a markedly different terrain. I took a course in Ethical Analysis (G. E. Moore up through Rawls), one course in Kantian Ethics, two courses on Rawls and his interlocutors (Nozick, MacIntyre, Sandel, etc.), and an unusual course on British Idealist Ethics. Along the way, I encountered in other courses treatments of the moral philosophies of Aristotle, Aquinas, Hume, Sidgwick, and Mill. In all of my courses, however, not a single word was ever uttered about any role for empirical considerations being relevant to moral theory, insofar as "empirical" included an important role for experimental inquiry. Aristotle, Hume, and Mill might have been recognized as naturalists of some sort, but we were not taught to think of them as basing their theories on natural science. The only reference to naturalism consisted in the emphatic reminder that we must *never, ever* commit the dreaded "naturalistic fallacy."

In contrast to this earlier dismissal of empirical considerations, as I write today there are stacks of books piling up on my desk faster than I can read them, all of which marshal empirical and experimental evidence in support of their basic claims. I attribute this radical transformation of focus to two developments: (1) evolutionary psychology got us

CHAPTER SIX

focused on the animal and human origins of morality, and (2) cognitive neuroscience turned our attention to how our bodies and brains, in interaction with our environments, make possible our moral experience, cognition, feeling, judgment, and reasoning. I regard this change as a very promising and much-needed development that will profoundly enrich future moral philosophy, by introducing into the discussion scientifically sophisticated accounts of human nature, motivation, deliberation, and judgment.

However, in spite of my conviction that moral philosophy must coevolve with empirical research on mind, thought, and values, I cannot regard all of naturalized ethics as an unqualified good. The chief problem is that some enthusiasts for the utilization of empirical research have drawn substantial conclusions about moral cognition that do not seem to me to be justified by the data and studies they cite. One especially disturbing trend in some of this recent work is the resurgence of the idea that all "normal" humans possess a moral faculty, moral instinct, or moral grammar that underlies intuitive judgments about what is morally right and wrong. One of the most influential recent statements of this view is formulated by Marc Hauser:

Moral judgments are mediated by an unconscious process, a hidden moral grammar that evaluates causes and consequences of our own and other's actions. . . . I show that by looking at our moral psychology as an instinct—an evolved capacity of all human minds that unconsciously and automatically generates judgments of right and wrong—that we can better understand why some of our behaviors and decisions will always be construed as unfair, permissible, or punishable, and why some situations will tempt us to sin in the face of sensibility handed down from law, religion, and education. (2006, 2)

Moral instinct talk is appealing to those who see it as a way to underwrite the idea of universal moral intuitions. They believe that nearly all humans rightly praise certain kinds of actions and rightly condemn others, and they conclude that such judgments are products of a universal moral instinct.

John Mikhail (2000, 2007, 2008) has also proposed a "universal moral grammar," and there are a number of other major books and articles that are currently debating the existence and workings of universal structures underlying moral cognition. My analysis and criticism focuses primarily on Marc Hauser's *Moral Minds: How Nature Designed Our Universal Sense of Right and Wrong* (2006), because it is one of the stronger, clearer, more carefully argued, and most influential contemporary versions of the moral instinct view—a book worthy of considered response.[2] I am not alone in criticizing Hauser's postulation of a moral organ. Ron Mallon (2008) has also provided a critique, but along different lines than I propose here.[3]

Although the idea of evolutionarily developed moral dispositions makes good sense (Hinde 2002; Flanagan and Williams 2010), I regard the positing of an alleged moral faculty, moral instinct, or moral grammar as a regressive move, in three fundamental respects. First, the assumption of a moral faculty is at odds with at least some recent empirical work on the nature of mind and thought. Second, it is not necessary to assume any such universal faculty in order to explain what might appear to be transcultural moral values and principles. Third, searching for a moral faculty tends to divert our attention from what is actually going on in human moral deliberation.

Postulation of a moral organ or instinct represents a twenty-first-century revival of a scientifically discredited eighteenth-century faculty psychology that tries to preserve the idea of universal cognitive structures defining right and wrong conduct. It is an attempt to embrace a naturalistic, scientifically grounded approach, while at the same time smuggling back in the idea of moral governance or guidance imposed on our experience from some allegedly universal source, this time in the form of universal moral instincts. Dewey (1922) articulated the major objections to the positing of universal instincts, and we would be well-advised to recall the lessons he taught us about the natural sources of human morality.

In short, my argument against the positing of a moral faculty will be that such a notion is scientifically questionable and constitutes a step backward in our attempt to understand human moral valuing and deliberation from a naturalistic perspective. It is not just that we *do not need* to assume a moral instinct, but rather that such an assumption is counterproductive to our efforts to understand moral values and judgments. I shall argue that we can preserve Hauser's (and others') most important empirically supported insights about shared dimensions of morality without committing ourselves to his unnecessary and misguided notion of a moral faculty or moral instinct. A naturalized perspective on the nature and sources of human morality is better served by not proposing some absolute foundation for our moral judgments, even if there do turn out to be some shared dimensions of moral systems (Flanagan and Williams 2010; Haidt 2012).

CHAPTER SIX

The Mistaken Analogy between Moral Sense and Chomskyan Grammaticality

The foundational assumption of "moral grammar" theories is that "there may be deep similarities between language and morality, including especially our innate competencies for these two domains of knowledge" (Hauser 2006, 37). Hauser's version is based John Rawls's claim that just as Chomsky supposedly discovered a capacity for syntax—a sort of "language instinct"—underlying grammaticality in all natural languages, so also is there a sense of justice, and by extension a moral sense (a "moral instinct"), that generates intuitive moral judgments within a culture. In *A Theory of Justice* (1971), John Rawls famously articulated this grammar/morality analogy as follows:

> Now one may think of moral philosophy at first (and I stress the provisional nature of this view) as the attempt to describe our moral capacity. . . .
> A useful comparison here is with the problem of describing the sense of grammaticalness that we have for the sentences of our native language. In this case the aim is to characterize the ability to recognize well-formed sentences by formulating clearly expressed principles which make the same discriminations as the native speaker. . . . A similar situation presumably holds in moral philosophy. (46–47)

The foundational idea for Chomskyan linguistics is that ordinary native speakers are able to produce grammatically correct utterances and make grammaticality judgments about expressions in natural languages because they share an innate universal language (syntax) faculty. Correspondingly, in the moral context, it is claimed that people from different cultures can recognize the same moral principles, values, or dispositions because they all share a universal moral faculty. Hauser sets out to describe this moral faculty and to show how it generates universal judgments about which types of actions are morally permissible, which are impermissible, and which are morally obligatory.

Chomsky argued that nobody could learn a natural language on the basis of empirical generalizations from what they had heard spoken or seen written. There were two prongs to his "poverty of the stimulus" argument: (1) What a child hears in spoken language is often ungrammatical, consists mostly of mere parts of whole grammatical utterances, and is peppered with starts, stops, interruptions, and changes in midsentence. (2) There are a potentially infinite number of grammatically acceptable sentences one could utter, and nobody is ever exposed, even over their entire lifetime, to more than a tiny fraction of them. Chomsky inferred from this experiential under-determination that there must be innate universal principles of grammar underlying all natural languages and that those principles are not learned by inductive generalization. He concluded that there must be an innate universal grammatical competence, possessed by humans, but not other animals, which is enacted in competent linguistic performance, even though the speaker or hearer is typically unable to articulate even the most basic syntactic principles of their native language. Grammatical differences across natural language families were to be explained as possible transformations of deep universal syntactic structures.

Hauser assumes that Chomsky's view is well established in contemporary linguistics and that something like it must be the correct account of our language capacities. Then, on the basis of an alleged language/morality analogy, he concludes that there should be universal moral principles, just as there are supposedly universal grammatical principles:

My explanation for these disparate observations [about widespread moral judgments] is that all humans are endowed with a *moral faculty*—a capacity that enables each individual to unconsciously and automatically evaluate a limitless variety of actions in terms of principles that dictate what is permissible, obligatory, or forbidden." (2006, 36)

Just like a native speaker's linguistic competence, "a mature individual's moral grammar enables him to unconsciously generate and comprehend a limitless range of permissible and obligatory actions within the native culture, to recognize violations when they arise, and to generate intuitions about punishable violations" (44).

One reason why many have found the moral grammar model (and the corresponding linguistic analogy) so appealing is that, if true, it would give one possible explanation for the existence of intuitive moral judgments. As we saw earlier, there is considerable evidence of at least two basic pathways of moral thinking—one a nonconscious, intuitive mode of cognition, and the other a conscious, rational, reflective mode. Recall Hauser's capsule summary of these two processes:

What is central to the discussion ahead is that intuition and conscious reasoning have different design specs. Intuitions are fast, automatic, involuntary, require little attention, appear early in development, are delivered in the absence of principled reasons, and often appear immune to counter-reasoning. Principled reasoning is slow, deliberate, thoughtful, requires considerable attention, appears late in development, justifiable, and open to carefully defended and principled counterclaims. (31)

As we have seen in earlier chapters, there is a growing body of experimental evidence supporting this idea that most of our moral judgment goes on beneath the level of conscious awareness, in more or less automatic processes of evaluation and response (Casebeer 2003; Haidt and Joseph 2004; Haidt 2012). One of the important implications of this two-track idea is that what most philosophers have traditionally taken to be the core of moral reasoning—namely, conscious, reflective analysis and deliberation—is actually mostly after-the-fact justificatory reflection that is not the motivating source of our moral judgments. In chapters 3, 4, and 5, I explained why this is a significant insight, but I then argued that any adequate account of moral cognition must also include a process of moral deliberation as imaginative dramatic rehearsal. But why think, as Hauser seems to, that the *only* way to explain the nonconscious, intuitive level of moral judgment is to postulate a more or less inborn moral faculty that generates our fast, unreflective judgments? I am proposing an alternative account that, instead of positing a moral instinct, borrows from the cognitive linguistic view that we learn language by learning patterns and constructions.

On the face of it, what is appealing to many theorists about this language/morality analogy is the way it brings the idea of strict moral governance into play. There are alleged to be cognitive structures, operating mostly at a nonconscious level, that supposedly give rise to specific judgments about the morality of certain actions. In other words, it is claimed that there are clear constraints *imposed* on us from sources deep within our minds (and brains). We do not *make* these principles; rather, we *discover* them as they generate both our unreflective and reflectively considered moral judgments alike. As I argued earlier, this is not God telling us what to do (as in traditional theological ethics), nor is it pure practical reason telling us what to do (as in Kantian rational ethics), but it *is something telling us what to do and what not to do*—something over which we supposedly have no control and that imposes right and wrong from deep within us.[4]

Problems with the Idea of a Grammatical Faculty or Instinct

Hauser's view is predicated on what he calls the "Linguistic Analogy" between the grammatical capacities underlying natural languages and the moral capacities underlying cultural moral systems (Dwyer, Huebner, and Hauser 2010). Let me state emphatically that I do not deny the usefulness of comparing linguistic communication with moral deliberation and judgment. There most certainly *are* important similarities between linguistic and moral competence. Here are five: (1) Both are interactive communicative and cooperative human practices involving notions of human flourishing; (2) both are learned behaviors; (3) both require shared human cognitive and affective capacities, some of which might reasonably be called "universal" and what McCauley (2011) calls "maturationally natural"; (4) both involve constraints governing their successful performance, some coming from the nature of our bodies and brains, and others coming from social and cultural practices and institutions; and (5) both involve what Paul Churchland (2007) calls "skills of learning" to navigate a complex social space.

My argument is not, therefore, against the linguistic analogy as such. Rather, I will be taking issue with Hauser's self-described "strong version" of the analogy that requires "a commitment to the way of studying the mind and cognitive capacities that has emerged since the beginning of the cognitive revolution over fifty years ago, spearheaded by, among others, generative linguists" (Dwyer, Huebner, and Hauser 2010, 487). What I *will* take issue with is Hauser's strong Chomskyan (generative linguistics) commitment to an innate instinct or faculty as the basis for some perceived behaviors (either linguistic or ethical). My argument will be that, even granting structural parallels between certain linguistic and moral phenomena, it does not follow that one must postulate either a linguistic or a moral instinct or faculty in order to explain the phenomena in the two respective domains (i.e., of language and morality). I will suggest that the moral phenomena that "moral grammar" theorists claim are explained only, or at least best, by a hidden moral faculty can be explained better by a naturalistic approach that makes no assumption of innate instincts. So, it is not the linguistic analogy as such that I take issue with, but only Hauser's strong Chomskyan interpretation of that analogy.

Hauser's first and most catastrophic error is his grounding of a language/morality analogy on a mistaken Chomskyan view of language. Hauser is not alone in assuming the correctness of Chomsky's views about the structure of a linguistic theory. In fact, almost everybody who posits a "moral grammar" appears to base it on Chomskyan assumptions about linguistic competence and performance (see, for example, Mikhail 2007; Mahlmann and Mikhail 2005; Sripada 2008; Harman 2008). Hauser and his coauthors repeatedly stress that what they are taking from Chomsky is *not* any of the particular details of Chomsky's linguistic principles; instead, what they appropriate is Chomsky's conception of the structure of a syntactic theory as positing an innate grammatical capacity. Thus Dwyer, Huebner, and Hauser say, "The human mind includes a biological mechanism that provides a limited range of possible moral systems, and on this basis, the environment selects among the cultural options on offer to acquire a particular morality" (2010, 492).

I will argue that, just as we do not need to assume an innate grammatical faculty to explain language, so likewise we do not need to assume a moral faculty to explain morality. There is not space here, nor is it my purpose, to give a detailed critique of Chomskyan linguistics. What I hope to do, instead, is to call into question Chomsky's conception of a linguistic theory as needing to describe an innate language (syntax) faculty that generates grammaticality judgments, since it is this kind of mechanism that Hauser is proposing to explain moral judgment. I can at least point out how Chomsky's view, which was once taken as gospel, has been subjected to withering criticism over the past forty years, and that there are now empirically supported alternative accounts of language learning that do not presuppose innate language modules or instincts.

The first thing to notice is that Chomsky's view is predicated on the rigid distinction between syntax, semantics, and pragmatics. Grammatical form is claimed to be independent of meaning, which is claimed to be independent of the uses to which sentences are put. Chomsky correctly saw that the only way to make sense of a pure language organ is to regard it as entirely concerned with generating grammatical form (syntax), uncontaminated by the meaning of what is being said (semantics) or the use to which that meaning is being put (pragmatics). This idea of pure form led to the hypothesis that syntax must be realized in neural modules in the brain, which take inputs and generate outputs, but which are more or less autonomous, in the sense that they are not influenced by their relations to multiple other brain regions.

However, recent theories of language that draw on contemporary cognitive neuroscience argue that brain modularity alone could never explain human language, which requires massive parallel processing across multiple interacting areas of the brain, including numerous cortical and subcortical regions that communicate via re-entrant neural mappings. Jerome Feldman, who constructs neurally based computational models for language processing, sums this up:

Current knowledge in the brain and behavioral sciences makes it clear that language, like all other human mental capacities, involves an intimate interaction between nature and nurture. We are not born a blank slate, nor is learning just selecting values for a few parameters. It also makes no biological sense to talk about an autonomous module for grammar or any other capability. The brain clearly does rely on specialized neural circuits, but these interact massively with one another and almost always have overlapping functions. . . .

If language is, as the neural theory suggests, continuous with other mental activities, it makes no sense to ask if certain evolutionary adaptations are specialized only for language. There is no such thing as language in isolation from thought. (2006, 282)

Feldman, George Lakoff, and many other linguistically oriented cognitive scientists are currently developing the branch of cognitive linguistics known as the *neural theory of language*, one of the chief empirically based alternatives to Chomskyan innatism. In sharp contrast to Chomsky's claims, the neural theory of language approach argues that form is *meaningful* because of the nature of our bodies, brains, environments, interests, values, and practices: "Grammar is inherently coupled to the form, meaning, and use of language" (Feldman 2006, 282). Feldman suggests that an embodied theory of language, such as the neural theory he espouses, rests on two basic principles:

1. Thought is structured neural activity.
2. Language is inseparable from thought and experience. (336)

And I would add a third principle, which I believe he would also accept:

3. Thought is shaped by the nature of our bodies, our brains, and the environments with which we interact in an ongoing fashion.

Neither syntax nor meaning in natural languages can be explained by neural modules alone, but must rather take account of complex interactive sensory, motor, and affective systems that are activated in our engagement with our multidimensional and frequently changing environments. To cite just one example, the neural theory of language argues that language acquisition involves the learning of *constructions*—recurring patterns encountered in spoken and written languages—in which phonetic, syntactic, semantic, and pragmatic dimensions are interwoven. In what is known as *embodied construction grammar* (Langacker 1987/1991; Lakoff 1987; Lakoff and Johnson 1999; Goldberg 1995, 2006; Feldman 2006), a construction is a pairing of linguistic form and meaning that consists of intricately linked schemas and other cognitive structures that arise from the nature of human bodies, brains, and the environments they interact with. Also, as we saw in the previous two chapters, conceptual meaning involves sensory and motor simulations (Bergen 2012).

For my purposes here, it is unnecessary to flesh out the neural theory of language. My point is only to suggest what an alternative non-Chomskyan view of language learning might involve and how it would be tied to the body, the brain, and the environment in a way that does not posit a language organ. I shall later suggest that such an empirically based non-innatist account of language acquisition can provide a naturalistic model for thinking about our moral competence, too, once we correctly understand the strong parallels between our learning of a language and our gradual acquisition of a moral system.

In short, the postulation of an innate language faculty is not necessary to explain language, even though there will obviously be some modular systems involved in our linguistic processing. Correspondingly, I contend that neither is it necessary to postulate an innate moral faculty in order to explain our intuitive moral judgments. Let us see why.

What's Wrong with Hauser's Conception of the "Moral Faculty"?

My key claim in this section can be stated simply as follows: There is no moral faculty, in any significant sense of that term. The form of my argument is also straightforward: A brief survey of even a few of the components that Hauser attributes to the "moral organ" is sufficient to show that, from the perspective of cognitive neuroscience, there could be no such organ as a distinct faculty. There are simply far too many complexly interacting multifunctional systems that have to be in play in our intuitive moral judgments for there to be anything remotely resembling a distinct moral faculty. My shorthand way of capturing this point is to say that Hauser's alleged "moral organ" could be nothing short of a full-fledged human being!

To make it clear that I am not caricaturing Hauser's position, here, in his own words, is a large portion of his explicit statement of what he means by our "moral faculty":

1. The moral faculty consists of a set of principles that guide our moral judgments but do not strictly determine how we act. The principles constitute the universal moral grammar, a signature of the species.

2. Each principle generates an automatic and rapid judgment concerning whether an act or event is morally permissible, obligatory, or forbidden.
3. The principles are inaccessible to conscious awareness.
4. The principles operate on experiences that are independent of their sensory origins, including imagined and perceived visual scenes, auditory events, and all forms of language—spoken, signed, and written.
5. The principles of the universal moral grammar are innate.
6. Acquiring the native moral system is fast and effortless, requiring little to no instruction. Experience with the native morality sets a series of parameters, giving birth to a specific moral system.

. . .

9. To function properly, the moral faculty must interface with other capacities of the mind (e.g., language, vision, memory, attention, emotion, beliefs), some unique to humans and some shared with other species. (2006, 53–54)

In order to see why the notion of a moral faculty is nearly vacuous, we need only consider even a partial list of the most basic components that Hauser regards as comprising this alleged faculty. The size and scope of this list is staggering, and the capacities contained therein in no way constitute a single discrete moral faculty. Not surprisingly, since our moral judgments are mostly directed at actions past, present, and future, large parts of the alleged moral faculty will involve our capacity to understand and assess actions, agents, and intentions. Hauser's list includes at least the following components of the moral organ (with page numbers indicating where he treats these in more detail):

1. *The ability to comprehend event structure and the logic of events* (171–78). Hauser lays out five principles by which we identify something as an action (as opposed to a mere event) that involves agent causality and goal-directedness.
2. *A narrative sense of self* (182–87). The attribution of action and moral responsibility requires an agent with at least a partially unified sense of self embedded within larger narrative frameworks.
3. *A capacity for empathy and a functioning emotional system* (187–96). We must be able to imagine the experience of others, take up something like their perspective, and respond emotionally. The human "mirror neuron" system would be central to our ability to empathize.
4. *Disgust reactions* (196–200). Some of our strongest moral responses are grounded in bodily disgust reactions (Rozin, Haidt, and McCauley 2008).
5. *Possession of a mature "theory of mind"* (200–206). We must understand that other people have mental states, perspectives, feelings, and intentions somewhat like, but often not identical with, our own.

6. *A capacity for categorization* (208–14). We must be able to recognize kinds of agents, kinds of actions, kinds of situations, and so on, in order to grasp the applicability of our moral principles.
7. *Patience* (214–19). We must have an ability to forestall immediate gratification for long-term satisfaction or values.

This is only a very partial list of systems and capacities necessary for moral judgment. Hauser acknowledges additional systems that would need to be included in a full account, such as perceptual abilities, memory, imagination, and so forth.

However, the previous list is quite sufficient to make my point, which is that *what we are describing are some of the many interwoven abilities that constitute a whole human person. These are capacities necessary for most of our larger-scope cognitive operations, and they depend on a more-or-less functioning brain, operating in a more-or-less functioning body (consisting of multiple interdependent bodily systems), permeated with affective valence, and inextricably intertwined with a physical, interpersonal, and cultural environment.* It makes no sense, and it does no useful work, to call this massive interconnection of systems and environmental affordances a distinct moral faculty.

Drawing on genetic theory and cognitive neuroscience, Patricia Churchland has challenged the idea that there might exist a specific genetic basis for anything like a "moral organ" with a determinate set of moral capacities. Hauser's postulation of an innate moral organ would require a distinct genetic basis for those capacities. Churchland argues, to the contrary, that "given evolution's *modus operandi*—tinkering-opportunistically rather than redesigning-from-scratch—we should not expect our conception of functional categories at the macro level to map neatly onto genes and gene products" (2011, 98–99). The reason for this is that "the relations between genes and behavior, as geneticist Ralph Greenspan observes, are not one-to-one, not even one-to-many; they are *many-to-many.* . . . By and large, the strategy of trying to link a single gene to a particular phenotype has been superseded by the understanding that genes often form networks, and that a given gene is likely to figure in many jobs" (97–99). Because most genes play a role in *multiple* networks that are involved in many different cognitive functions and behaviors, Hauser's innateness hypothesis is highly problematic:

The basic lesson then is that working backwards from the existence of a certain behavior to a brain region that supports that behavior to the innateness of a function is, especially in animals that are prodigious learners, a project fraught with evidential hazard. . . . The classical problem that bedevils all innateness theories of behavior is that in the absence of supporting evidence concerning genes and their relation to brain circuitry involved, the theories totter over when pushed. (Churchland 2011, 109, 116)

Churchland acknowledges that if one day it should turn out that there are completely universal moral themes, principles, or behaviors (which she doubts), then the idea of a moral organ might be one possible explanation. However, it is simply a mistake to think that only a species-specific moral faculty could guarantee shared moral judgments. She suggests that a far more plausible explanation would be that our moral reasoning recruits preexisting cognitive and affective capacities as a sophisticated form of problem-solving. In other words, this widely displayed behavior "may also just be a common solution to a very common problem" (107). Instead of our having evolved a distinct moral instinct or faculty, it is far more likely that we have recruited a large number of preexisting systems for addressing the kinds of problems we tend to call "moral problems," and that over the millennia we have cultivated various strategies for dealing with the kinds of problems (moral and otherwise) that we have tended to encounter on a more or less regular basis, given our nature as currently evolved and the kinds of environments we find ourselves in.

Other prominent neuroscientists have offered similar arguments against any notion of a moral faculty of the sort proposed by Hauser. The partial list of capacities that Hauser regards as requisite for moral judgment (given above) cannot be usefully localized to any unique, or even distinct, set of functional neural assemblies or regions of the brain. As any neuroscientist will confirm, the massively parallel processes of ordinary human cognition and feeling do not form anything like a distinct set of faculties or organs. The rule is not exclusive modularity, but rather interconnection and re-entrant circuits among multiple brain areas. What current neuroscience evidence argues for is a combination of modularity and widespread integration of neuronal assemblies. As Gerald Edelman puts it: "We now know that modularity of this [rigidly autonomous] kind is indefensible. The alternative picture, that the brain operates only as a whole [the holistic view], will also not stand up to scrutiny.... The long-standing argument between localizationists and holists dissolves if one considers how the functionally segregated regions of the brain are connected as a complex system in an intricate but integrated fashion" (2004, 30–31).

Don Tucker, who studies the functional architecture of the brain, argues for massive interconnection of brain regions that involves both the more functionally differentiated cortical shell areas and the core limbic structures involved in motivation and emotion:

CHAPTER SIX

What we find are amazingly intricate networks, each embedded within the other and each linked to its neighbor by highly stereotyped connections. . . .

At the core must be the most integrative representations, formed through the fusion of many elements through the dense web of interconnection. . . . Meaning is rich and deep, with elements fused in a holistic matrix of information charged with visceral significance. Emanating outward from this core neuropsychological lattice are the progressive articulations of neocortical networks. Finally, at the shell are the most differentiated of networks, differentiated internally with the finest of the network architectures in sensory cortices. . . . (2007, 179–80)

On the basis of functional brain structure architecture, Tucker is arguing that although there are some more or less modular structures in the sensory and motor cortices, these structures are connected to areas in the limbic core that process experience in a more holistic, emotional, and intuitive fashion. In anything as complex as moral judgment, typically all of the systems listed by Hauser are going to be active together in integrated networks. Moreover, *the multiple systems involved will not be uniquely dedicated to making moral judgments*. What Antonio Damasio says about human reason would apply equally to our capacity for moral judgment:

Human reason depends on several brain systems, working in concert across many levels of neuronal organization, rather than on a single brain center. Both "high-level" and "low-level" brain centers, from the prefrontal cortices to the hypothalamus and brain stem, cooperate in the making of reason.

The lower levels in the neural edifice of reason are the same ones that regulate the processing of emotions and feelings, along with the body functions necessary for an organism's survival. In turn, these lower levels maintain direct and mutual relationships with virtually every bodily organ, thus placing the body directly within the chain of operations that generate the highest reaches of reasoning, decision making, and, by extension, social behavior and creativity. (1994, xiii)

In his more recent work that is primarily focused on moral cognition, Damasio explicitly applies this general claim about the complex interactions of different levels of cognitive activity to deny the existence of a core moral faculty:

Not surprisingly, I believe that ethical behaviors depend on the workings of certain brain systems. But the systems are not centers—we do not have one or a few "moral centers." Not even the ventromedial prefrontal cortex should be conceived as a center. Moreover, the systems that support ethical behaviors are probably not dedicated to ethics exclusively. They are dedicated to biological regulation, memory, decision making, and creativity. Ethical behaviors are the wonderful and most useful side effects of those other activities. But I see no moral center in the brain, and not even a moral system, as such. (2003, 165)

In short, the idea of a moral faculty is scientifically suspect. It is a gussied-up twenty-first-century holdover from Enlightenment faculty psychology that needs to be jettisoned in any cognitively realistic theory of mind, thought, language, value, and action. The original idea behind faculty psychology is that human thought can be divided into a small set of *discrete types* of cognitive judgments—scientific, moral, technical, political, aesthetic, religious, and so on. Each of these types of judgment is then alleged to be generated by the distinctive activity of one or more mental faculties, such as sensation, understanding, reason, feeling, imagination, and will.

The foundational assumption of faculty psychology—that thinking consists of discrete, autonomous kinds of judgments—seems to be accepted by Hauser when he begins his book with the claim that certain types of judgments are distinctly and uniquely "moral" in character. Recall from chapter 1 the example he constructs of a young girl whose father gets angry at her for hitting a playmate and then later gets angry again when she puts sand in her mouth. How, he asks, does that child know that her father's first disapproval (of her hitting her playmate) was *moral* in character, while his second disapproval (of her eating sand) was only *prudential* (here, a matter of hygiene)? In Hauser's words, "How do the child's emotions send one action to the moral sense and the other to common sense?" (2006, 30).

Notice that this question would only make sense if you thought, as Hauser does, that there must be two different and independent senses (one moral and the other prudential) operating in the case of the little girl. But, as I argued earlier, why should we think that we have an autonomous *moral* sense that is distinct from our *aesthetic* sense, which is, in turn, different from our *prudential* sense, and so on? Our experience does not come pre-partitioned into types—moral, aesthetic, religious, prudential, technical—that require distinct and independent types of judgment.[5] My point here is that, if we reject the pre-partitioning of kinds of experiences (or judgments), we will not be so inclined to think that there must be a moral instinct or faculty underlying what we think of as uniquely *moral* judgments, since we will no longer postulate uniquely moral judgments.

At this point I can imagine a defender of a moral grammar claiming that I have unfairly caricatured Hauser's view when I suggest that his "moral faculty" must be some kind of relatively encapsulated neural module or cluster of modules. If he is not committed to some sort of modularity thesis, they might argue, then all of my talk about the massively parallel character of cognitive processing might seem to be beside the point. Hauser might simply embrace the profound complexity and multidimensionality of mind by identifying the alleged moral faculty as a particular specific cluster of cognitive capacities working together.

There are at least two problems with such a reply. The first is that Hauser does seem to embrace some sort of encapsulation of capacities when he says, "If our moral faculty can be characterized by a universal moral grammar, consisting of a set of innately specified and inaccessible principles for building a possible moral system, then this leads to specific experiments concerning the moral acquisition device, its relative encapsulation from other faculties, and the ways in which exposure to the relevant moral data sets particular parameters" (Hauser, Young, and Cushman 2008, 125). "Relative encapsulation" sounds like modularity talk to me.

The second problem is more devastating, and it is my key criticism. If there is no moral module, then what sense does it make to posit a moral faculty? What significant work is moral faculty talk doing here, other than to give a name to a large number of highly complex functional systems that have to be operative for us to make moral judgments?[6] If these many systems are not themselves dedicated only to moral cognition, but instead serve a large number of different functions in different contexts, then how do they form a "faculty"? As Flanagan has observed:

> Despite the appeal of thinking of moral psychology as somewhat modularized, a problem arises in thinking of the various psychological competencies involved in moral responsiveness as *too* separate, distinct, and modular. . . . It is possible that their [the dispositions and competencies] boundaries are determined as much by the interest-relative concerns of those of us who find the analysis of such things useful as they are by natural psychological borders. Furthermore, to whatever extent the distinctions among psychological competencies are based on natural demarcations, the fact remains that most of the putative dispositions and competencies are thought to interact with one another in complex ways and to have permeable borders. (1991, 240)

It might be pointed out that Flanagan (Flanagan and Williams 2010) has recently argued for some version of moral modularity. However, this is consistent with the above quotation and with a denial of a moral faculty. By "moral modules," Flanagan and Williams mean "evolutionarily ancient, fast-acting, automatic reactions to particular sociomoral experiences" (430), and they cite Haidt's five (now six) moral foundations as examples. They claim to remain agnostic about Hauser's attempt to reduce all these modules to one supermodule. I would only suggest that, as I have argued above, this would rob the notion of a moral module of any distinctive or unique content, since it would have to include an extremely broad range of capacities and dispositions, most of which are involved in cognitive activities that stretch far beyond moral judgments.

In short, faculty psychology is as old as Western philosophy and came into special prominence in the Enlightenment, before we had much of a scientific understanding of cognition, as a way of marking out alleged types of judgment and explaining them by the operations of distinct faculties. Today, however, we have made considerable progress in delving beneath the positing of discrete faculties and exploring the vast interaction of organic systems required for the maintenance of various types of organism. Faculty talk has ceased to do important explanatory work.

The Illusion of a Moral Instinct or Faculty

The point I am trying to make here was made some ninety years ago in Dewey's withering critique of the tendency to posit universal instincts as a way of explaining what appear to be universal human behaviors. Dewey's argument was simply that hypothesizing universal instincts is just bad science, insofar as instinct talk takes a vast complexity of determining factors, all subject to changes in context, and reduces them to a single "instinct" or causal source. He begins by acknowledging how tempting it is to explain why we harm each other by attributing it to some universal instinct of aggression, or to explain our helping someone as caused by some universal instinct of human sympathy (or love, or respect for others), but Dewey then argues that the many sources and "causes" of various human behaviors are complex, multilayered, and profoundly context-dependent. Positing deep human "instincts" amounts to trying to reduce all of this motivational complexity to a single homogenous *cause*. Dewey concludes:

In spite of what has been said, it will be asserted that there are definite, independent, original instincts which manifest themselves in specific acts in a one-to-one correspondence. Fear, it will be said, is a reality, and so is anger, and rivalry, and love of mastery of others, and self-abasement, maternal love, sexual desire, gregariousness and envy, and each has its own appropriate deed as a result. Of course they are realities. So are suction, rusting of metals, thunder and lightning and lighter-than-air flying machines.

But science and invention did not get on as long as men indulged in the notion of special forces to account for such phenomena. . . .

After it had dawned upon inquirers that their alleged causal forces were only names which condensed into a duplicate form a variety of complex occurrences, they set about breaking up phenomena into minute detail and searching for correlations, that is, for elements in other gross phenomena which also varied. (1922, 104)

The core of Dewey's argument, when applied to the moral cognition, is that causation (physical, economic, social, moral, or of whatever kind) cannot be adequately understood by reducing the complexity of situations and actions to some alleged unitary faculty or instinct. Again, this kind of reductionism amounts to no more than giving a single name to what is actually a vast interaction of complex functional systems. For example, the causes of war, Dewey says, cannot be reduced to some universal instinct of bellicosity—some allegedly primitive warlike character possessed by all humans. Positing an instinct of bellicosity (or aggression) does no useful explanatory work. That kind of reductionism, especially in the realm of human morality, politics, and international affairs, has only led to disaster—to war and more war—because it fails to understand the complexity of the causes and motivations of various human behaviors. It explains almost nothing, and offers no ameliorative guidance, to assert that the cause of war is some warring instinct instilled deep within every human breast.

What Edward Wilson says about the many types of aggression, which cannot be reduced to a core aggressive instinct, could be applied to other alleged instincts:

Like so many other forms of behavior and "instinct," aggression in any given species is actually an ill-defined array of different responses with separate controls in the nervous system. No fewer than seven categories can be distinguished: the defense and conquest of territory, the assertion of dominance within well-organized groups, sexual aggression, acts of hostility by which weaning is terminated, aggression against prey, defensive counterattacks against predators, and moralistic and disciplinary aggression used to enforce the rules of society. (1978, 101–2)

Granted, we must mark distinctions and recognize patterns in experience, if we want to understand how and why things happen, but such discriminations are only selective abstractions from the complex of realities operating in a given situation. As Dewey saw it, "the tendency to forget the office of distinctions and classifications, and to take them as marking things in themselves, is the current fallacy of scientific specialism. It is . . . the essence of false abstractionism. This attitude which once flourished in physical science now governs theorizing about human nature. Man has been resolved into a definite collection of primary instincts which may be numbered, catalogued and exhaustively described one by one" (1922, 92).

In short, instinct and faculty talk is overly abstractive and does no important explanatory work. Instead, we need to jettison faculty talk and focus on the best empirical research on human cognition, much of which can be found in the works of the moral faculty theorists I am criticizing.

The Paucity of Evidence for a Universal Moral Grammar

I have been claiming that the idea of a moral instinct, supposedly underlying a moral grammar, is scientifically suspect. It is unsound because instinct and faculty talk is unsound. Understandably, proponents of the moral grammar faculty have attempted to muster relevant empirical evidence to support their claims. Therefore, I want to take a passing look at some of the kinds of evidence that are assembled in support of moral instinct claims. Moral grammar theorists tend to point to a universal (or quasi-universal) set of principles or dispositions underlying human moral judgments, which they then attribute to the moral faculty that generates these universals. To the contrary, I will argue that even if there should turn out to be a few universal dispositions, values, or principles (which is somewhat doubtful), those phenomena could be explained by means other than positing a moral grammar.

Here are some of the major types of evidence that are typically used to support moral faculty arguments. John Mikhail (2007) lists four types of alleged evidence for a universal moral grammar: (1) Even young children appear to have the basics of a well-developed legal code. They learn how to parse actions with respect to issues of intent, they distinguish "moral" from "social convention" issues, and they recognize that false factual beliefs, but not false moral beliefs, can be exculpatory. (2) All natural languages seem to have terms for key moral (deontic) concepts such as *permissible, impermissible, obligatory*. (3) There appear to be universal transcultural prohibitions of certain actions such as murder, rape, and undeserved harm to innocents. (4) Certain brain regions seem to be activated for moral deliberation and judgment.

Even more recently, Susan Dwyer, Bryce Huebner, and Marc Hauser (2010) have also attempted to list supporting scientific evidence for the linguistic analogy between a linguistic and a moral grammar. They cite three main types of evidence. The first comes from "Trolley-type" moral dilemmas where subjects are asked to make decisions about cases in which one must weigh harms to one person or group in relation to some other person or group, in circumstances in which *some* harm cannot be avoided and under varying conditions of direct versus indirect causation of the harm. A growing number of studies suggest that there may be some cross-cultural consensus concerning general principles governing decisions of this sort. Interestingly, though, the authors acknowledge, with admirable candor, that the interpretation of these results remains unclear: "Such data highlight some of the complexities involved in our moral competence. Although moral judgments display a pattern suggestive of principled distinctions (across individuals), and independent of many matters of content, the fact that various factors interact in important ways with one another has yet to be fully appreciated" (Dwyer, Huebner, and Hauser 2010, 497).

A second source of evidence comes from studies of how people appear to calculate the magnitude of harms. Noting that current research indicates at least three cognitive mechanisms for calculating magnitude, one unpublished study by Huebner, Miller, Seyedsayamdost, and Hauser observes that people tend to use only one of those three available mechanisms in their utilitarian moral calculations. Most people appear to believe that, regardless of the total number of people harmed, an action is judged permissible (or even obligatory) if it redirects a lethal threat in a way that saves even one more person than is lost. Thus, it would be judged permissible to kill 500 people to save 501. The authors conclude that "these results set up the hypothesis that judgments of moral permissibility in the context of utilitarian calculations are largely impenetrable to the numerical calculations that are carried out by the two core nonlinguistic systems for enumeration" (Dwyer, Heubner, and Hauser 2010, 498). In other words, moral judgments based on calculations of moral harm appear to use but one principal mechanism of the three that are available, suggesting that only one mechanism has been selected for moral calculations of harm.

A third kind of evidence, based on Heubner and Hauser's unpublished data, suggests that certain types of moral reasoning show immunity-to-context effects. Whereas one might expect that the order of presentation of cases or examples might well affect a person's judgment of a particular case, Huebner and Hauser (unpublished data) examined whether the order and context for presenting cases had any significant effect on people's judgments concerning classic moral dilemmas that are the standard focus of much contemporary research. They found that the order did not seem to matter and concluded that "these data suggest that folk-moral judgment—and hence the computations that underwrite it—is predominantly impenetrable to domain-general heuristic strategies" (Dwyer, Huebner, and Hauser 2010, 500).

I shall not discuss all of the empirical studies that are cited in these linguistic analogy ("moral grammar") debates to support claims for a distinct moral faculty. However, the few that I have surveyed above are representative, and they are sufficient to make my point that *none of these "experimental" results lead necessarily to the conclusion of a moral grammar.* It is true that if there were such a thing as a moral faculty, it would explain these types of evidence. The argument would run as follows: These apparently universal principles or dispositions or values or judgments could be the result of the workings of a moral faculty shared by all normal humans. *My counterclaim is that there are alternative explanations for any evidence of nearly universal principles, capacities, cognitive mechanisms, or values. Therefore, even if there should turn out to be any moral universals, there is no necessary connection between that fact and the positing of a moral faculty.* As I will argue later, there are other ways to explain (1) the types of evidence I have just surveyed, and (2) how such generalities might be found across cultures and come to be learned by individual members as they grow and develop to adulthood.

Other "Moral Grammar" Views

I have focused my criticism on the very strong form of the moral grammar theory, which claims that it is universal moral *principles* that issue from the moral faculty. Some have argued that weaker versions—versions that do not insist only on universal principles—might not be subject to the kinds of criticisms I have brought against Hauser. Consequently, there is a growing recent literature focused primarily on what the term "universal" is supposed to range over. For example, Chandra Sripada (2008) has nicely summarized three popular specifications of what might count as "native" and "universal."

(1) *The simple innateness model* claims that there are *universal moral principles and rules* possessed by "normal" adult humans in all cultures. Sripada counters by observing that there are just too many exceptions found in various cultures to support any strong allegation of universal principles. Gilbert Harman (2008) has replied to Sripada that denying that there are many such norms does not mean that there are not at least *some.* However, I find it revealing that when Harman comes to listing

candidates for universal principles, he does not actually supply a list, but instead reverts to little more than observing that various moral philosophers throughout history have cited the principle of double effect as central to our moral reasoning.

(2) *The principles and parameters model* proposes that there exist some very *general abstract moral principles that contain parameters* that might be specified differently within different cultural settings. For instance, one could imagine a very general rule against harming members of a certain protected class of persons C, but different cultural groups might define membership in the protected class C differently, thereby giving rise to noncongruent prohibitions across cultures concerning who may and who may not be harmed. The obvious appeal of this type of approach is the opportunity to hold on to universal norms while being able to explain where and why variations arise. Hauser, Young, and Cushman thus claim, "Our moral faculty is equipped with a universal set of principles, with each culture setting up particular exceptions by means of tweaking the relevant parameters" (2008, 122). Hauser is thus a principles and parameters theorist.

Sripada observes that this second model is typically put forward as the solution to the alleged problem that there is a gap between the complex learning target of moral principles (and parameters) and the actual resources available to any person who supposedly learns these moral principles. The idea of innate principles and parameters is typically offered as the proper explanation of how a person could ever come to full adult moral understanding when the "data" they experience amounts to no more than what Kant famously called "a disgusting mishmash of patchwork observations and half-reasoned principles in which shallowpates revel" (1785/1983, 410). Sripada criticizes this model for assuming an analogy between a linguistic grammar and a moral grammar on the issue of the poverty of the stimulus. The heart of Sripada's counterargument is that contrary to what we find in the case of language learning, where there is a true gap between the linguistic performance we are exposed to and the linguistic competence we manifest, in the case of moral learning there does not seem to be a serious poverty-of-the-stimulus gap between what a person is exposed to in everyday life and the ideal competence that is supposed to be manifest in mature moral agents.

(3) The *innate biases model* asserts that all humans possess certain innate predispositions (such as disgust reactions) that make the presence of certain moral norms more likely under certain circumstances. On this view, the universal moral organ would be a set of innate "preferences, aversions, emotions, and other elements of one's psychology" (Sripada

2008, 333) that incline that person (or group) to affirm certain types of moral principles. According to this model, it is the psychological predispositions that are universal, while general principles (perhaps universal, perhaps not) are the likely result of these dispositions.

Sripada thus shows that there are stronger and weaker views about what is universal in human nature, and he favors the weakest (innate biases/moral disposition) view as more compatible with the empirical evidence about variable cultural values and practices. This is a reasonable view, supported by a good deal of research into culturally variable moral systems. There is a growing consensus that, at the very least, there are a small number of more-or-less universal human predispositions to certain types of behaviors (e.g., see Hinde 2002; Flanagan and Williams 2010; Haidt 2012). These predispositions have emerged over our evolutionary history, so that from our current perspective they present themselves as tendencies of humans to approve of certain values and social arrangements and to disapprove of others. However, considerable caution is advisable regarding any strong claims to universality. Jesse Prinz (2008) has done a superb job of marshaling relevant counterevidence against many of the most popular current claims about innate moral structures, whether in the form of universal rules, universal moral domains, or even a supposedly universal distinction between moral and non-moral concerns. Prinz's counterevidence raises serious doubts about whether there really are any substantial moral universals of any sort.

Yet, even if we grant, for the sake of argument, the apparent universality of a very few moral dispositions, domains, principles, or values, it would *not* necessarily follow from this assumption that there must be a moral faculty (with a moral grammar), or that only a moral faculty could explain that universality. My contention throughout has been that there is no convincing scientific evidence of a universal innate moral faculty, and there is plenty of scientific evidence against such a faculty. Moreover, the key question would still be whether we can explain the emergence and transmission of purported generally shared biases, abstract norms, or specific principles without reference to a moral faculty. I am arguing that we can.

Morality without a Moral Faculty

I observed earlier that the idea of a moral faculty has historically been connected to the belief that experiences and judgments come pre-differentiated into discrete types—theoretical, technical, moral, aesthetic, and so on—with each such judgment type being the product of the activity of one or more distinct faculties. Thus, Hauser claims that there are distinctly moral experiences, giving rise to distinctively moral judgments, which result from the operation of our moral faculty.

I have argued that we ought to jettison this assumption of a distinctly moral experience (with its hypothesized correlative judgment type), because neither our experience nor our judgments are neatly precategorized as segmentable into types. As I argued in chapter 1, there is no more an exclusively "religious" experience than there is an exclusively "moral," "political," "scientific," or "aesthetic" experience. Instead, these names mark aspects of the rich and thick multidimensionality of any given experience. Sometimes in our inquiries, we may happen to find one selected aspect more important in light of our present situation, interests, and values, and so we focus exclusively on that dimension. For example, calling an experience "moral" is characteristically our way of highlighting the fact that in some measure its outcome and development pertains to well-being. Nevertheless, there are typically aesthetic, political, technical, and scientific dimensions, among others, tied up in any developing experience of well-being. As soon as we hypostatize any selected quality of a situation and then proceed to define the entire situation by that abstracted quality, we overlook most of the complexity and depth of that situation and the moral judgments we subsequently make about it.

It is long past time to abandon the conviction that there are distinctively moral situations that require uniquely moral judgments issuing from an innate moral organ. If situations don't come pre-stamped as "moral" or "aesthetic" or "technical," then it becomes far less tempting to go searching for some innate moral faculty to generate distinctively *moral* judgments.

I have argued that we should see *morality as a form of complex problem-solving*—the reworking of a situation that has become problematic and has inhibited our ability to skillfully, meaningfully, and harmoniously navigate our social space. We could then recognize all of the capacities and operations that Hauser and others have identified (under their mistaken notion of the moral faculty or moral grammar) as relevant to our ability to negotiate our moral space, and we could then treat them for what they are—namely, *not* a unified moral organ, but a constellation of human capacities and propensities for making sense of our experience and engaging in problem-solving forms of inquiry.

There is no need to postulate a moral faculty to explain the fact that cultures around the world and throughout history show a predilection for certain very general values, and perhaps even for some very abstract moral principles. Instead of a single moral faculty, there appear to be some basic dimensions that characterize moral systems around the world. Shweder, Much, Mahapatra, and Pack (1997) identify three major categories (autonomy, community, divinity), while Haidt (2012) argues for six (care/harm, fairness/cheating, loyalty/betrayal, authority/subversion, sanctity/degradation, liberty/oppression). In addition to these dimensions of moralities, there are also, as we have seen, various catalogues of basic virtues and taxonomies of values. Flanagan and Williams have called Haidt's five dimensions of moral systems "moral modules" that are evolutionarily established and show up in moralities around the world and throughout history. They claim that "moral competence consists of, or is the emergent product of, a set of autonomous or relatively autonomous sociomoral competences" (2010, 431).

I do not think "modularity" is the best word for these dimensions of moral systems, because, as I have said above, I do not see such dimensions as suitably encapsulated to constitute true modules. At any rate, it should be clear that I take no issue with various taxonomies of recurrent aspects of moral systems. I agree that there are recurrent dimensions across cultures, but nothing about them requires reference to discrete instincts, organs, or even distinct modules (in the narrow sense of that term that posits encapsulation). Rather, they are simply the result of the fact that human beings have evolved with certain capacities, needs, dispositions, and forms of human interaction that make it unremarkable that cultural moral systems the world over would recognize at least a few nearly universal moral concerns, values, and principles, even if there is sometimes cultural variation in the interpretation of those principles.

Finally, as I mentioned above, if Hauser had relinquished his reliance on a moral faculty, most of the empirical research he nicely summarizes could be put to the service of a naturalized moral theory, to the extent that it represents reliable cognitive science evidence about human moral experience, cognition, and judgment. We could even retain the language/morality analogy, but only on the condition that we adopt something akin to an embodied construction grammar view of meaning and language, instead of a Chomskyan innatist perspective. As we have seen, according to the neural theory of language, people learn language by learning basic constructions and how those recurring patterns of form, meaning, and use are combined and blended in various contexts.

The parallel to moral development is obvious: we learn skillful coping strategies within complex social-moral spaces of communal interaction. If our development is successful, we learn both what the basic moral constructions (values, ends, virtues, practices, institutions) are, and we also learn appropriate forms of criticism, analysis, argument, justification, and communication within these moral spaces. There is nothing eternal, pure, or radically transcendent about any of this—no ultimate ground of value. There is, instead, the fact of never-ending moral inquiry, carried out under conditions of very limited knowledge, often in times of great stress and turmoil. In such cases, as Dewey (1922) sagely observed, moral deliberation is more like the making and judging of art than it is a matter of discerning and following absolute rules or universal principles. Moreover, just as there is no universal human "art faculty," neither is there any universal "moral faculty"—and neither do we need such a faculty.

SEVEN

Moral Fundamentalism Is Immoral

I have suggested that the greatest sin of moral philosophy is moral fundamentalism, by which I mean the belief in unmediated access to absolute, universal foundational moral truths. Moral fundamentalism typically takes one or both of two forms: either it assumes that we have access to absolute moral principles, or it assumes that we have access to foundational moral facts (about what is good, right, virtuous, etc.). Fundamentalism is a sin because it (1) attempts to reduce the relevant complexity of human experience to simple abstractions, (2) denies the human necessity for interpretation, and (3) shuts off moral inquiry. These are three of the worst things a person can do when it comes to engaging in moral deliberation.

In this chapter I want to explain, first, why moral fundamentalism is a non-starter when viewed from the perspective of the best empirical scientific research on how the mind works, and, second, why moral fundamentalism is immoral. My point is to guard against the kind of backsliding I examined in the previous chapter, where I criticized moral instinct theories as one form of moral fundamentalism. My contention in this chapter is that our desire for certainty and the avoidance of doubt—our flight from our fallibility—is so strong that we find it devilishly difficult to avoid regressing to absolutist thinking, even though it is irresponsible from both a scientific and moral perspective to do so. Just as we saw with the recent proliferation of moral faculty (or moral instinct) theories, I will now argue that the recent popularity of moral realism (of a certain foundationalist kind) represents a regression to an epistemologically untenable and morally irresponsible standpoint. I will challenge the belief in absolute moral principles and the correlative idea that humans have access to a bedrock of distinctively moral facts.

The Nearly Irresistible Lure of Fundamentalism

By moral fundamentalism, I mean the doctrine that there exist foundational moral truths that take the form of either absolute, unconditional, universally binding moral laws or a set of absolute and foundational moral facts. The ideal of universal, unconditional moral laws is so appealing because it satisfies our deep human desire for an ultimate order and moral framework for our world. After all, who wouldn't like to have access to unshakable moral truths by which to unwaveringly guide their moral decisions? Who wouldn't feel better knowing that there are moral absolutes and that humans can grasp them through revelation, reason, or intuition?

Our longing for an absolute and transcendent moral order runs deep in the human psyche. In his essay "Religion as a Cultural System," Clifford Geertz attempts to explain the virtually universal appearance of religious systems throughout human history and in all cultures as a response to fundamental human needs for stable meaning, order, and morality. He defines religion as

(1) a system of symbols which acts to (2) establish powerful, pervasive, and long-lasting moods and motivations in men by (3) formulating conceptions of a general order of existence and (4) clothing these conceptions with such an aura of factuality that (5) the moods and motivations seem uniquely realistic. (1966, 4)

My focus here is not on religion as such, but instead on the fact that we humans seem to have a primitive need for some kind of order in our world, an order that we can make at least some sense of. To be fully satisfying for most people, that order must be both rational and moral. Otherwise, we feel lost and adrift in a universe in which we are not at home. Geertz observes that humans do not necessarily need to know precisely what that overarching order is, just so long as they can maintain confidence that there *is* such an order. What people cannot abide, however, is the idea that our constructed symbol systems are incapable of giving us at least some access to the absolute order of things, an order that would allow us to manage our worldly affairs:

> The thing we seem least able to tolerate is a threat to our powers of conception, a suggestion that our ability to create, grasp, and use symbols may fail us, for were this to happen we would be more helpless, as I have already pointed out, than the beavers. The extreme generality, diffuseness, and variability of man's innate (i.e. genetically programmed) response capacities means that without the assistance of cultural patterns he would be functionally incomplete, not merely a talented ape who had, like some under-privileged child, unfortunately been prevented from realizing his full potentialities, but a kind of formless monster with neither sense of direction nor power of self-control, a chaos of spasmodic impulses and vague emotions. (Geertz 1966, 13)

What Geertz so elegantly captures is our profound human anxiety over the possibility that we might be cast adrift in the cosmos without the necessary symbolic resources to guide us truly and rightly—that is, without the symbolic capacity to discern some order of things and to morally align ourselves with that perceived order.[1] Our deepest human concern is not just a desire for some cosmic ontological order within which to find our place, but especially a profound desire for an overarching *moral* order—a conception of the universe as governed by principles of right action that, if followed, will ultimately (if not in this world, then certainly in the next) lead to justice and deserved happiness.

In *The Future of an Illusion* (1927), Freud reduced this human longing for a moral order to the child's (and later the child grown to adult's) profound psychological need for protection through love and the assurance of ultimate justice:

> As we already know, the terrifying impression of helplessness in childhood aroused the need for protection—for protection through love—which was provided by the father; and the recognition that this helplessness lasts throughout life made it necessary to cling to the existence of a father, but this time a more powerful one. Thus the benevolent rule of a divine Providence allays our fear of the dangers of life; the establishment of a moral world-order ensures the fulfillment of the demands of justice, which have so often remained unfulfilled in human civilization; and the prolongation of earthly existence in a future life provides the local and temporal framework in which these wish-fulfillments shall take place. (1927/1968, 31)

I am concerned here not so much with Freud's uncharitable take on religious belief, but rather with his insight into the strong psychological

CHAPTER SEVEN

force of our human desire for a moral world order. It is just such a desire for moral absolutes that makes moral law views so appealing, because they claim to provide us either with an all-powerful and all-knowing moral authority (God), a surrogate moral authority in the form of universal reason (as in Kant's system), or a metaphorical conception of society as a person who issues moral commands (Lakoff and Johnson 1999, chap. 14). And although Kant insisted that morality cannot be grounded on divine commandments, he nevertheless went on to argue (in what has come to be known as the "moral" argument for the existence of God) that only the existence of an all-powerful and moral God could insure that morally correct actions would eventually (in another world in the afterlife) be rewarded with a deserved happiness not typically achieved in our finite, earthly existence.

Sartre, too, stated with admirable clarity, in "Existentialism Is a Humanism," that any loss of conviction in the existence of moral absolutes grounded in some universal essence would indeed leave humanity in an uncomfortable state of anxiety, moral anguish, and despair. For, if there is no pre-given, absolute, universal moral guidance available, then we are abandoned to our own human devices to construct whatever moral order we can manage—a moral order without transcendent foundations.

Utilizing a form of argument reminiscent of certain Kantian claims about our subjection to absolute (categorical) imperatives, C. S. Lewis famously asserted that the undeniable existence of a moral order to which we are inescapably bound, which he thought all normal folks would experience as binding on them, is compelling evidence of the existence of a supreme moral lawgiver. In his influential radio programs aired weekly in 1942, during some of the darkest hours of World War II, and later published as *Mere Christianity* (1952/2001), Lewis's argument for an overarching moral order that gives rise to unconditionally binding moral laws gave great comfort and hope to his fellow citizens who were then confronting the horrors of aerial bombardment. In a nutshell, Lewis's argument runs as follows:

1. All normal (i.e., non-sociopathic) humans have a sense of being morally bound by laws of right action that indicate right versus wrong behaviors. We all feel the call to practice decent behavior, even as we know that we often fail to do so: "We are forced to believe in a real Right and Wrong. People may be sometimes mistaken about them, just as people sometimes get their sums wrong; but they are not a matter of mere taste and opinion any more than the multiplication table" (7).
2. "The Law of Human Nature, or of Right and Wrong, must be something above and beyond the actual facts of human behavior. In this case, besides the actual facts,

you have something else—a real law which we did not invent and which we know we ought to obey" (21).
3. The compelling power of Moral Law over me shows itself to me as an inner commandment, imposed on me by something beyond and infinitely more powerful than me (24).
4. These moral commandments must be issuances of a "Being behind the universe (who) is intensely interested in right conduct" (30) and is "more like a mind than it is like anything else" (30).
5. "If the universe is not governed by an absolute goodness, then all our efforts are in the long run hopeless. But if it is, then we are making ourselves enemies to that goodness every day . . ." (31).

I have oversimplified Lewis's argument, though I think I have captured the gist of it. When a friend some time back urged me to read *Mere Christianity* so that we could discuss it, I remember being stunned by what an embarrassingly bad argument I discovered there. It seemed to me that any college freshman ought to be able to see what's wrong with this form of reasoning. For, even if we were to grant (for the sake of argument) a felt sense of being bound by standards of right behavior, there are any number of plausible explanations of such a feeling that do not lead to the postulation of a divine moral being as the source of that moral constraint. To cite just three plausible alternative sources, the feeling of being morally bound might issue from (1) evolutionary development, (2) parental expectations, or (3) societally inculcated demands for cooperative behavior and submission to authority. Lewis would have claimed, I suspect, that none of these finite sources could account for the *absolute* bindingness of moral obligations that assert their demands on us even over against parental and societal norms. However, that argument would only work if you assumed that a person would not feel morally bound or obligated unless the source of the obligation was an absolute (and nonnatural) authority. That is akin to saying that we cannot have a thought or a feeling of the absolute, the infinite, or the unconditional, unless the source of that thought or feeling is something actually absolute, infinite, or unconditional—and that is simply false.[2]

Setting aside Lewis's shockingly weak arguments for a divine lawgiver, I want to focus, instead, on his alleged *fact* that we are bound by absolute moral constraints and thus inhabit a *moral universe governed by moral laws*, without which all would be chaos and ruin (steps 1 and 2 of his argument above). Such a universe, if it exists, might be more comforting than a world where there is no overarching moral framework for all humankind, and this, no doubt, is why Lewis's view proved so important for the lives

CHAPTER SEVEN

of so many Christians and even some non-Christians during and after World War II, capturing their profound desire for an essentially moral, or morally guided, universe.

In other words, what is most telling is not so much the positing of a supreme moral agent to explain our felt obligations, but rather the strength of the need for absolute moral laws, or at least an absolute moral order, that is the basis for the entire discussion. Why should anyone think that there are but two choices, namely, either we are subject to unconditional moral laws, or else morality is a made-up fiction? Why should any sense of moral obligation that a person might experience have to be absolute and unconditional in order for it to play an important, binding role in our lives?

The fear of rampant relativism lurks everywhere behind such assumptions about moral obligations. The idea that, absent an *absolute* moral order that supplies unconditional moral constraints, anything and everything is permitted seems terrifying enough to tempt many into the arms of moral fundamentalism, back to the comfort of moral absolutes imposed from beyond our finite human condition. Geertz sums up this gnawing anxiety:

The strange opacity of certain empirical events, the dumb senselessness of intense or inexorable pain, and the enigmatic unaccountability of gross iniquity all raise the uncomfortable suspicion that perhaps the world, and hence man's life in the world, has no genuine order at all—no empirical regularity, no emotional form, no moral coherence. And the religious response to this suspicion is in each case the same: the formulation, by means of symbols, of an image of such a genuine order of the world which will account for, and even celebrate, the perceived ambiguities, puzzles, and paradoxes in human experience. (1966, 23)

I am suggesting that the urgent and persistent lure of moral fundamentalism is grounded in (1) our profound anxiety that there might not be an overarching moral order, and (2) our misguided sense that *only* some absolute, transcendent source of values, moral order, and moral laws could supply the moral guidance we so desperately desire. One contemporary manifestation of this profound moral anxiety is the extreme anger and hostility rampant in various subcultures in American society (and across the globe) that is directed toward anyone who questions any part of their moral orthodoxy. They regard any challenge to one of their primary moral convictions (e.g., such as their views on homosexuality, constraints on firearm possession, or taxation) as a fundamental challenge to the very existence of morality itself! They seem to believe that if even one of their moral values or principles is shaky, then their entire moral edifice could come tumbling down.

My response to this seemingly unbreakable attachment to moral absolutes will be twofold: (1) that, as a consequence of how human conceptual systems work, humans do not have access to absolute moral knowledge, and (2) that we most certainly do not need any such foundational or absolute grounding for our moral convictions in order to have a moral society. I turn first to a criticism of absolutist views of moral principles, and then to claims about the existence and self-evidence of so-called "moral facts."

Moral Fundamentalism Is Based on an Erroneous Conception of Mind and Thought

Two decades ago, in *Moral Imagination: Implications of Cognitive Science for Ethics* (Johnson 1993), I set out my argument against one version of moral fundamentalism that I christened the Moral Law folk theory. The basic idea behind this particular version of moral fundamentalism is that there are universal moral laws that unconditionally bind human beings, by virtue of their being creatures possessed of reason. One popular version of the theory sees God as the source of these moral principles, while the non-theological rationalist version takes the principles or laws to issue from a universal reason. Here is a list of some of the basic underlying assumptions upon which the Moral Law folk theory rests:

1. *Universal Reason:* Every person—every moral agent—possesses a capacity for reason that is capable of functioning according to logical principles. In most versions, this rationality is taken to be an essential feature that distinguishes humans from other animals.
2. *Universal Moral Laws:* There is a determinate set of universal moral principles, discernible by human understanding and reason, binding on all rational creatures, and jointly circumscribing all our fundamental moral duties and prohibitions.
3. *Absolute Moral Values/Laws:* Moral values, ends, and laws are absolute, and therefore are not contingent on any historical or cultural conditions or circumstances.
4. *Moral Concepts Are Literal:* If our fundamental moral laws (principles) are to be absolute and definitive of our duties, then the concepts contained in those principles must be highly determinate and well defined; otherwise, they would provide no definite and unconditional moral guidance. These concepts must map directly on

to states of affairs in the world and must be univocal in their meaning, lest we be left with moral indeterminacy. In other words, these basic moral concepts must be literal.
5. *Classical Category Structure:* The univocal meaning of these concepts must be defined by sets of necessary and sufficient conditions for their application to experience. A particular situation falls under a moral principle only if it satisfies the conditions defined by the concepts used in the moral principle.
6. *Hierarchy of Principles, Ends, Values:* Universal reason must supply principles for rank-ordering ends, values, and secondary principles, in order to avoid conflicts of duties and indeterminacy about which moral ends should take priority in a given situation.
7. *Radical Freedom:* Fundamentalism assumes that moral agents are capable of acting on principles that they recognize as binding, and it assumes that such totally free choices are what make possible the attribution of moral responsibility for one's actions.

My first contention about moral fundamentalism is that an overwhelming body of empirical evidence from the cognitive sciences shows that each of the above seven claims is false—not just a little bit false, but badly false, at least as they are formulated within fundamentalism. I think it is no accident that fundamentalists tend to believe in disembodied mind, because in order for fundamentalism to work, you would need a disembodied, decontextualized reason, concepts acquired *a priori*, and a pure logic that is utterly independent of our physical and social experience.

What I have said in previous chapters and that earlier book (Johnson 1993) calls into question some of these basic tenets of the Moral Law folk theory. However, instead of rehearsing again my arguments against each of these seven tenets, I will focus instead on just one of those tenets that serves as a linchpin for the whole orientation of moral fundamentalism—namely, the idea that our moral concepts (e.g., *person, murder, lie, promise, right*) must be pre-given, literal, fixed, and highly determinate, in order for our moral reasoning to apply to concrete cases in experience.

Let us take, as an example, the pivotal concept of moral personhood. It would be difficult to imagine a more foundational concept for morality in American and European culture, where morality is construed as centered on the idea that, as creatures possessing moral personhood, we have moral obligations to treat all other creatures possessing moral personhood in certain clearly specified ways and to refrain from treating them in certain prohibited ways. Conversely, others have a duty to treat *us* (as possessing moral personhood) as we are required to treat them. In other words, it is your status as a *person* that both imposes moral responsibilities on you and also entitles you to specific moral consideration from others. For instance, among the range of principles proposed by some theorists as moral universals, at the top of virtually everyone's list would be commands like *"It is morally impermissible to kill an innocent person."* A principle of this sort is frequently cited as the basis for absolute prohibitions on murder and abortion (assuming that you can make the case that a fetus is a person). Moreover, if you remove the "innocent" restriction, then this principle can perhaps be broadened to prohibit not just murder and abortion, but also capital punishment, killing in warfare, and euthanasia.

Let us, therefore, consider what moral fundamentalism requires concerning the character of the concept *person*, and then ask whether human concepts actually work in this required fashion. First, the concept of moral personhood would need to be what Lakoff (1987) calls a "classically structured category," that is, it must be definable strictly by a set of necessary and sufficient conditions for membership in the category. If we are to apply the concept properly to a particular situation, there must be a discrete and clearly definable set of features, properties, or characteristics the possession of which makes an individual or state of affairs in the world a member of the category in question. For instance, if you believe that you have an absolute obligation to respect anything that is a person, then you could only follow this command if you knew how to determine precisely who is a person and who is not. Alan Donagan, who offers a Kantian-style rationalist moral theory centered on the concept of respect for moral personhood, thus recognizes the necessity of a strict, fixed definition of personhood (as rational creature) when he confidently asserts:

The fundamental concept of respecting every human being as a rational creature is fuzzy at the edges in the superficial sense that its applications to this or that species of case can be disputed. But among those who share in the life of a culture in which the Hebrew-Christian moral tradition is accepted, the concept is in large measure understood in itself; and it is connected with numerous applications, as to the different weights of which there is some measure of agreement. (1977, 71)

The key phrase here is "the concept is in large measure understood in itself." I submit that this statement is dramatically false, and its falsity is hardly lessened to any serious degree by adding "among those who share in the life of a culture in which the Hebrew-Christian moral tradition is accepted." If Donagan were correct, then reflective, sincere, morally earnest Christians and Jews would agree on abortion, capital punishment, war, and euthanasia—to name but a few key issues—because they would all be able to grasp what the concept *person* means, and they would all be able to determine what the concept of respect for persons (as rational creatures) requires of us.

However, quite obviously these are today issues of vehement disagreement among various Jewish and Christian sects. The idea that there is, as Donagan assumes, *the* Hebrew-Christian moral tradition, in any sense that would guarantee unanimity of judgment on such all-important moral issues, strikes me as unsupported by the historical record, which reveals instead multiple developing traditions, in the sense articulated by Alasdair MacIntyre in *After Virtue* (1984). If the concept of respect for moral personhood were "understood in itself," then most assuredly the morality of abortion, to cite just one key issue, would be definitively settled among Jews and Christians, *which it most certainly is not!* Donagan seems implicitly to understand that, as MacIntyre argues, our best chance of shared moral values would require our shared participation in a fairly homogenous moral tradition that is not undergoing change in light of changing historical circumstances. Maybe then, for a short period of time, there might be sufficient cultural unanimity and shared values to render certain moral concepts and principles temporarily free of significant contestation. It is no accident that MacIntyre cited medieval monastic communities as social arrangements that were so stable (I had almost written "stagnant") as to insure near fixity of beliefs and practices for extended periods of time, in relative isolation from changing conditions in the broader society. Only under such conditions of resistance to change could one have the slightest hope of there being univocal foundational moral concepts grasped equally by all. However, as MacIntyre has shown, moral traditions have histories and often change in the face of changing material, social, and cultural conditions. Donagan's idea of a *single* "Hebrew-Christian moral tradition" is a fiction, and a historically changing fiction at that.

In other words, the only way that a concept could be "understood in itself" would be if the concept were so determinate in its defining features, and so immune to any historical and cultural changes, that it univocally picked out a fixed set of cases to which it applied, completely independent of any traditions of inquiry and practice. I can imagine someone who held such a view supporting their position by comparing the concept *person* to the concept *triangle*, which they take to be a truly universal concept determinately and unambiguously definable independent of any context. However, this alleged analogy between the concept *person* and the concept *triangle* will not work in the way that it is intended, because the supposedly literal concept *triangle* turns out to vary in its meaning relative to the particular geometry it is defined within, and there can be a number of different, often incompatible, geometries. That is, even mathematical concepts are context-, tradition-, and paradigm-dependent. This last qualification—that of specifying the relevant conceptual framework for any concept—is necessary, because there is no definition of *triangle* that applies univocally in every conceivable geometry. Mathematical concepts are no more context independent and absolute than our moral concepts, or any of our other concepts, for that matter (Lakoff and Núñez 2000).

The whole idea of a concept "understood in itself" is nearly useless, since all concepts are inter-defined relative to other related concepts within semantic fields that operate within various language games (Wittgenstein 1953), paradigms (Kuhn 1970), conceptual systems (Putnam 1981), idealized cognitive models (Lakoff 1987), and frames (Fillmore 1985; Lakoff 1992; Feldman 2006), all of which have histories and change over time. No concept is understood "in itself." Concepts get their meanings in relation to other concepts and to historical situations and frameworks in which the concept develops and gets its temporary application. Concepts grow and change as changing conditions require us to find new ways of ordering and making sense of our experience.

In order to see this contingency of our conceptual systems, let us delve more deeply into the concept of personhood. The concept of personhood that is used in moral theories and moral systems turns out to be somewhat specialized. It is not the same as *personality*, which is typically defined as the sum total of all the behavioral and mental (cognitive, emotional, temperamental) characteristics by means of which an individual is recognized as being the unique "person" that they are. Instead, "*moral* personhood" or "*moral* personality" is conceived as a feature or set of features that jointly constitute certain creatures as subjects having moral responsibilities toward other persons and, in turn, as worthy of like moral treatment from other creatures who are moral agents (i.e., have moral personhood). Throughout our long history of ascription of moral personhood to certain beings, when it comes to identifying what it is that makes some creature a locus of moral personhood, the one perennial answer is *rationality*—the possession of reason. Although in some theological frameworks personhood might be based on possession of a "soul," when asked how we know who has a soul, the answer typically reverts to some specification of rational capacity.

This pushes the question of personhood back to the question of what *rationality* consists in. It can be quite distressing to discover how nearly impossible it is to identify any clear folk theory of rationality as undergirding morality, because hardly anyone can articulate what they mean by saying that something or someone is rational. The best accounts usually fall back on equating being rational with being logical, but this only pushes the question back even further to what it means to be logical, which tends to leave out some of the very dimensions of reason that are most important for morality. I honestly doubt that one in a thousand ordinary folks could tell you with any clarity and specificity what constitutes a person's being possessed of reason, and I would observe that those who fall back on some weak notion such as "capacity to engage in logical thinking" would find it impossible to explain how possession of such a logical capacity has any bearing whatsoever on their status as bearer of moral duties and privileges!

Therefore, absent any clear folk theory of rationality, we are forced to turn to the explicit philosophical and theological accounts of rationality that pretend to explain why and how being rational is the key to morality. Students in introductory ethics courses routinely learn the mantra that "*reason* (or being rational) is what separates humans from brute (nonrational) animals and establishes them as moral agents," but few can go beyond that. It is perhaps Kant, aided by over two centuries of interpretation and commentary by professional philosophers, who makes the clearest connection between the possession of reason and moral personhood. As Thomas Hill Jr. (1980) explains, what Kant holds as definitive of moral personality is a cluster of capacities that are all allegedly manifestations of our capacity to reason. These include

- the capacity to act on principles or maxims (to act for reasons);
- the capacity to follow rational principles of prudence and efficiency (i.e., the capacity for hypothetical reasoning);
- an ability to foresee future consequences, adopt long-range goals, and resist immediate temptation (which presupposes a freedom of choice lacking in brute animals);
- the capacity and disposition to follow categorical imperatives (i.e. unconditionally binding moral laws);
- an ability to understand the world and to reason abstractly.

This is a very specialized list of rational capacities designed explicitly to support the claims Kant makes that only rational creatures are capable of moral judgment. We need not quibble over whether this list is either complete or completely adequate. I cite it only as illustrative of the kinds of capacities that philosophers have identified as connecting rationality to moral personhood, for here it is clear that the capacity to set ends for oneself, to grasp imperatives involved in practical reasoning, and to order one's behavior according to principles of right action are clear requirements for many traditional rationalist conceptions of morality. In other words, one can at least see how the possession of "rational" capacities of the sort listed above might be deemed necessary for a person to be a moral agent, whereas the commonplace that to be rational is to be logical reveals no such connection.

For the sake of argument, and as an act of charity, let us assume that something like this list is supposed to constitute the necessary and sufficient conditions for the attribution of moral personhood. Recall that, from the perspective of moral fundamentalism, the point of having such a list is to permit us to determine precisely and unequivocally who has moral personhood, so that we know to whom our moral principles apply, and how they apply. A prototypical instantiation of this sort of moral personhood would be

1. an adult human (assuming we could define "human") animal,
2. who had proceeded successfully through a "normal" process of human cognitive and emotional development,
3. thereby acquiring the cognitive and conative capacities listed above (e.g., the capacity to set ends, to deliberate about means, to reason abstractly, to inhibit one's bodily actions, etc.), and
4. currently capable of exercising all of these capacities.

Such an individual would instantiate what is known as a cognitive prototype (Rosch 1973, 1977) and would constitute a "central member" of the category *person*. If we held that all and only "persons" were worthy of moral treatment, then we would immediately recognize such a person as worthy of respect. Such a prototypical person would come as close as we could get to satisfying our idealized cognitive model (Lakoff 1987) of moral personhood, as defined roughly by Hill's criteria listed above.

So far, so good, the moral fundamentalist might say. Unfortunately, however, this case that satisfies Hill's criteria is only one central member of the complex radial category (Lakoff 1987) *person*. In no moral theory of which I am aware is this type of individual the only one to whom moral personhood is attributed. Here's why: if this cognitive model (based on the criteria specified in 1–4 above) were taken to be the univocal, literal definition of person, then *no infants, no aging adults who have lost some or all of the list of person-defining attributes of rationality, and no disabled individuals lacking one or more of these capacities would be accorded the status of moral personhood.* Babies, some children, certain disabled persons, and many elderly adults would lack any moral status, precisely because they do not fully satisfy the conditions for moral personhood, as defined by the central cognitive model that is based on the cognitive prototype given above.

The obvious problem is thus how to include certain beings who do not satisfy conditions 1–4 of the central cognitive model. The common strategic move in response to this obvious problem is to add a codicil of the following sort to the definition:

Any creature having the capacity to acquire the requisite rational abilities, and any creature who once had, but has now lost one or more such capacities, shall be awarded the status of personhood.

But why? Why should having once been rational entitle you to current treatment as a rational agent, when you have ceased to be one? If the *real* criterion is possession and exercise of rational capacities, then why should you get the benefit of moral treatment when you can no longer exercise such capacities? Moreover, why should some creature who might in the future, under certain "normal" developmental conditions, acquire rational capacities, be entitled to moral treatment? In other words, why should infants, who are not yet rational in the sense specified by the cognitive prototype, be entitled to respect? We might feel warmheartedly inclined to grant them such a status, but it is unclear that there is any conclusive argument for doing so.

However, even if we accepted the potential-to-be-rational extension of the concept, we would still be left with cases of infants born with massive cognitive and/or affective deficits that would place them outside the category *rational agent*, and even outside the category *creature who might become a rational agent* because they will never develop the ability to exercise the full range of rational capacities. For instance, a recent PBS television program on how people deal with death and dying related the heartrending story of a couple whose child was born with a rare degenerative condition in which he would never develop most, if any, of the list of rational capacities for moral agency listed earlier, and his condition would deteriorate until his death at some point in his second to third year of life. The child's remarkable parents showered him with loving care, afforded him all of the considerations of moral personhood, and shepherded him gently through the process of dying. Surely, we might think, everyone would admire these parents and say that they were right to act as they did. However, virtually none of this moral treatment could have been founded on a classically defined category of moral personhood, since the child lacked most or all of the rational capacities supposedly definitive of moral personhood and he was incapable of ever acquiring them.

In short, the category of moral personhood is not a classical category defined by necessary and sufficient conditions. It is a very complex radial category. According to radial-category theory (Lakoff 1987), typical human categories have a complex internal structure in which some members are "central" to the category (as specified by prototypicality effects), with other non-central members radiating out at different "distances" from the prototypical members. These non-central members will not share at least some of the conditions that define the central-member idealized cognitive model of *moral personhood*. For any candidates for possible membership in the category, we must make *evaluative (normative) judgments* about whether they should be included under the category and, if so, where they are to be placed in the cognitive geography of the category (i.e., how far from the center and how close to other non-central members). Notice that these judgments, although educated by familiarity with the radial category structure, are not made on the basis of having a set of concept-defining rules. Moreover, such judgments most certainly do not follow the rules that might be derived from, or defining for, the cognitive prototype.

The upshot of my previous analysis of *moral personhood* is that it is most definitely not a univocal literal concept in the sense that would be required for the Moral Law folk theory to work. The classical objectivist theory of literal concepts upon which moral fundamentalism is based is psychologically, cognitively, and linguistically unsound (Lakoff 1987) and cannot be fixed up. *Moral personhood* is typical of our most important moral concepts, insofar as it is a highly contested concept, has been so for many centuries, and shows every indication of continuing to be so in centuries to come. Quite simply, it is a changing concept, relative to historical contingencies. In American culture there were, obviously, times and places where African American slaves were not afforded the privileged status of moral personhood. Moreover, women have not been afforded full personhood at various times and places, and this holds even today in some cultures. Even more indeterminate is the status of so-called "brute" animals. Many environmental ethicists are recently arguing for the extension of moral personhood to certain non-human animal species, which would then entitle them to respect and moral consideration. An increasingly large number of people are coming to believe that certain animal species are the proper subjects of moral consideration usually afforded only to humans. The point is that decisions about such extensions cannot be generated by an algorithm operating on the cognitive model that describes the prototypical central member of the category.

An even bigger stretch, but one that is being argued for in some ecological circles, is the possibility of extending the notion of personhood to ecosystems, watersheds, and in general our more-than-human world, toward which we would then have moral obligations. Another arena of animated discussion concerns whether artificial intelligence systems achieve personhood once they are capable of performing "rational" calculations. Our newspapers and magazines are filled with discussions of whether artificial intelligence systems might someday merit the same moral consideration we today give only to certain members of one particular species (*Homo sapiens*). Finally (at least for now), we have begun what promises to be a long and contentious debate about whether to grant the status of personhood to corporations. In spite of the intuitive appeal of the Occupy Wall Street slogan that "only people are persons," the issue of corporations-as-persons is igniting a heated argument that is unlikely to be settled merely by appeal to shared intuitions.

To summarize: the concept *person* (or *moral personhood*), which lies at the heart of virtually every Western moral system and theory, is a complexly structured radial category that is subject to reconfiguration in response to changing historical circumstances, including changes in both material and cultural conditions. Deciding, at a particular point in time, who or what counts as a person is a matter of inquiry and requires evaluative judgment, subject to changes in our developing situation. Donagan's claim that such a concept could be "understood in itself" is, when taken at face value, ludicrous. A more charitable interpretation of Donagan's claim would be that, relative to a given historically developing moral tradition that is subject to fundamental change over time, some concepts achieve a level of temporary stability so that they can *seem* to be literal, univocal, clearly defined, and "understood in themselves." However, that most certainly does not make them absolute literal concepts in any sense of that term that would underwrite moral fundamentalism.

I would note in passing that things would get very much more complicated and far less determinate, were we to do a comparable analysis of the concept of personhood in traditions beyond Judaism and Christianity. If the open-endedness and relative indeterminacy of the concept is evident even within the history of a particular religious tradition (or cluster of traditions) like Christianity, as soon as we engage non-Western and various indigenous traditions, we get substantially different notions of personhood. In some forms of Buddhism, for example, this diversity might even extend to the denial of any unified self to be the bearer of personhood. There is no possibility whatever for finding a univocal, literal, universal, timeless concept of moral personhood.

My central point here is that moral fundamentalism cannot get off the ground without a foundation of classical category structure, and that theory has been demolished over the past thirty years (Wittgenstein 1953; Lakoff 1987; Gibbs 2006). Most of our key moral concepts are not literal, but are instead defined by multiple conceptual metaphors (Johnson 1993; Lakoff and Johnson 1999). For each of Haidt's (2012) six foundations of moral systems, it is possible to identify the conceptual metaphors that align with and provide our understanding of each foundation. The care foundation is understood primarily in terms of nurturance metaphors, which have their grounding in the nurturance of parents for their children. The fairness foundation gets articulated via what we called the "moral accounting" metaphor, in which good deeds earn moral credit whereas bad deeds (deeds that harm others) establish a moral debt. Within this framework, justice is a proper balance of moral accounts, in which each party gets what they deserve. The loyalty foundation is grounded on metaphors of bonding and binding, usually within a family group. Authority structures are metaphorically conceived within a "strict father" model of the family, in which the father is an authority who commands obedience, rewards good actions, and punishes bad ones. Models of the family then get extended metaphorically to humanity as a whole, giving rise to a universal ethics that has a built-in authority structure. Sanctity considerations are, as one would expect, tied to metaphors of purity and cleanliness. The liberty foundation is worked out through an extensive system of metaphors for freedom (Lakoff 2006) that have source domains involving the conditions for and restrictions on bodily motion in space. If a deed is metaphorically conceived as motion toward some goal-destination, then anything that blocks such motion is a metaphorical obstacle to freedom. Instead of univocal literal moral concepts, we thus found that our most basic conceptions were defined by multiple, often inconsistent metaphors, such as when moral action is understood metaphorically as nurturance versus when it is understood as maintaining or achieving purity.

In short, fundamentalism is based on a fundamentally mistaken view of concepts and the reasoning we do with those concepts. It is a cognitively unsupported and erroneous view of how conceptualization works. It is unable to give us cognitively realistic guidance for how we should think about morality and how we should understand moral judgment. It is scientifically unsupportable. I will later say more about why it is also immoral, but for now I will only observe that it sets up erroneous expectations about how we should resolve our moral problems, and it leads us away from serious moral reflection by closing off proper moral inquiry. That, I will argue later, is immoral.

Where Moral Realism Goes Astray

A frequent accompaniment to belief in absolute moral principles is a belief in foundational moral facts. In fact, when people claim to identify moral facts, those "facts" are quite often expressed as moral principles (such as, "It is wrong to do X"). Moral realism is the doctrine that there are such moral facts and that humans have cognitive access to them, as a basis for correct moral action. Just as there are several versions of moral faculty theories, there are varieties of moral realism. Because my challenge to moral realism rests on the very notion of a moral fact, I will not need to address the specifics of each of the different versions popular today. I have selected for consideration Russ Schafer-Landau's *Moral Realism: A Defence* (2003), because, like Hauser's position on a moral grammar, Schafer-Landau presents a version of moral realism that is clear, bold, and sophisticated in its arguments. On key ontological and epistemological issues, he is quite willing to hold fast to some very strong claims about the objectivity of moral principles and facts, as is evident in the following summary statement:

> I defend the view that moral principles and facts are objective in a quite strong sense: they are true and exist independently of what any human being, no matter his or her perspective, thinks of them. Moral facts are not scientific ones, and do not pass the standard test of ontological credibility; moral facts cannot be confirmed in the manner of scientific facts, but are no worse for that. Our moral beliefs are capable of motivating us all by themselves, and usually, but not invariably, do so. Moral obligations constitute reasons for everyone to act as they direct, regardless of whether such reasons bear any relation to one's existing commitments. . . . Despite skeptical worries, moral truths are knowable. At least some moral principles are knowable via self-evidence; moral conclusions are knowable if they are true and have been reliably produced. (8)

This passage is a concise summary of the key conclusions of the entire book. The central idea is that there exist "objective," "true," and independent moral principles and facts, upon which we can base a universal morality, even though there may be a plurality of moral systems compatible with this universal core of principles and facts. The point of stressing objectivity and facts is supposed to be that humans do not create morality, but instead discover it as a set of foundational facts and principles. The appeal of such an absolutist framework is obvious, for it would, if true, provide universal, objective constraints on moral behavior, thus avoiding any type of perspectivalism or relativism.

The idea of a "moral fact" is often used to summon up a possible analogy between natural scientific fact-finding, on the one hand, and some supposedly parallel activity of a moral science that discovers and explains *moral* facts, on the other. However, Schafer-Landau dismisses the form of ethical naturalism that would reduce moral theory to an empirical scientific undertaking, for he regards moral facts as constituting an independent, autonomous realm:

As I see it, there are genuine features of our world that remain forever outside the purview of the natural sciences. Moral facts are such features. They introduce an element of normativity that cannot be captured in the records of the natural sciences. They tell us what we *ought* to do; how we *should* behave; what is *worth* pursuing; what *reasons* we have; what is *justifiable* and what not. There is no science that can inform us of such things. (4)

According to this view, there are scientific facts and moral facts, and the latter have a normative dimension that is absent in the former. If these alleged foundational moral facts are not known by methods of inquiry standardly utilized in the natural sciences (i.e., experimental methods that are open to some measure of falsification and that seek some support from reproducible experimental evidence), then how are they known?

Having eschewed scientific methods (or any form of empirical inquiry, for that matter), Schafer-Landau is forced to embrace the notion of self-evidence. He proposes a "hybrid moral epistemology" that includes both claims about how our moral principles can be justified (a meta-ethical concern) and also concrete moral verdicts (a normative concern). As part of the former, he asserts that "at least some moral principles are self-evident: if one understands their content, and comes to believe them on this basis, then one is justified in one's moral belief" (8). Notice how close this claim is to Donagan's assertion that certain moral concepts or principles are "understood in themselves."

Schafer-Landau's position has strong affinities with G. E. Moore's infamous non-naturalism, to which he devotes part of an early chapter.

CHAPTER SEVEN

Recall, from my introduction, Moore's claim that there exists a *sui generis* moral property—*good*—that just happens to apply to certain states of affairs in the world. Moore argued that because the property *good* is non-natural, it cannot be analyzed into a set of natural properties and is, therefore, indefinable. Moore's claim is that we somehow just do grasp that a certain determinate state of affairs in the world is good, and that is all that can be said about it. We cannot explain why it is good, for to do so would commit the naturalistic fallacy, by attempting to cash out the concept in terms of natural properties.

Schafer-Landau's view is remarkably close to Moore's, although not focused merely on the concept *good*. It is a non-naturalism, defined as "one that insists on the non-identity of moral and *descriptive* properties, while allowing the moral to be entirely and exhaustively constituted by the descriptive" (76). The distinctive feature of Schafer-Landau's non-naturalism is the claim that moral properties are supervenient on natural properties, even though not reducible to any set of natural properties:

> According to the sort of ethical non-naturalism that I favour, a moral fact supervenes on a particular concatenation of descriptive facts just because these facts realize the moral property in question. Moral facts necessarily covary with descriptive ones because moral properties are always realized exclusively by descriptive ones. . . . Now, by hypothesis, the feature that is supposed to generate the different moral assessments *registers no effect at all on the descriptive world*. (77; italics added)

I have highlighted (in italics) the striking claim that nothing in the world, no observable state of affairs, is any different by virtue of its possessing or instantiating any non-natural property! Two *operationally indistinguishable* states of affairs (e.g., two actions) that have the same empirical description could support two *different moral* descriptions, according to Schafer-Landau. In other words, one could not determine from an empirical description of an action whether it is moral or not. This is precisely what Kant claimed in the *Grounding for the Metaphysics of Morals* when he asserted that "there is absolutely no possibility by means of experience to make out with complete certainty a single case in which the maxim of an action that may in other respects conform to duty has rested solely on moral grounds . . ." (1785/1983, 407).

Claims of this sort about the autonomy and independence of moral ascriptions from any empirical description strike me as saying nothing more than "certain actions have moral properties alighting on them, and that is all that can be said." How is this any different from the assertion that, given the requisite moral sensitivity and acuity, you will know a moral fact when you see one? I am reminded of G. J. Warnock's scathing criticism of Moore's claim that *good* is a non-natural property, indefinable, and apparently unrelated to any natural state of the world. Warnock writes sarcastically that, according to Moore's view, we are supposedly able to grasp "a vast corpus of moral facts about the world—known, but we cannot say how: related to other features of the world, but we cannot explain in what way: overwhelmingly important for our conduct, but we cannot say why" (1967, 16). It is difficult to imagine how this view of supervenience amounts to more than the confident affirmation that certain moral properties are just given—just out there (somewhere)—to be grasped as self-evident. Should you happen not to grasp the self-evidence of those moral facts, then so much the worse for *you*.

Self-Evident Truth Is a Bogus Concept

Schafer-Landau attempts to ground the objectivity and perspective-independence of moral facts on a strong notion of self-evidence. By a *self-evident* proposition, Schafer-Landau seems to mean any proposition "such that, once understood, one is justified in one's belief" (2003, 248). The idea is supposedly that, just by understanding the meaning of the proposition, one grasps its truth or falsity. Let us consider what might count, according to Schafer-Landau, as a relevant self-evident moral fact of the kind he is arguing for. There are surprisingly few of these alleged facts cited anywhere in his 322-page book. The best statement I can find is the following:

It seems to me self-evident that, other things equal, it is wrong to take pleasure in another's pain, to taunt and threaten the vulnerable, to prosecute and punish those known to be innocent, and to sell another's secrets solely for personal gain. When I say such things, I mean that once one really understands these principles (including the *ceteris paribus* clause), one doesn't need to infer them from one's other beliefs in order to be justified in thinking them true.... I don't know how to argue for the self-evidence of these (or other) propositions. (248)

The notion of self-evidence has a long and ignominious history as the precious darling of many foundationalist epistemologies and absolutist doctrines. It is unfortunately a common fallback notion for those who want to affirm foundational propositions when they have no arguments to defend their claims. If future historians of philosophy ever acknowledge any philosophical progress in the twentieth century, surely one clear instance of progress they will cite are the challenges to absolute foundations for knowledge that have come from virtually every philosophical orientation, from empiricist epistemology to hermeneutic phenomenology. Who—after Nietzsche, Dewey, Quine, Kuhn, Hanson, Feyerabend, Foucault, Lyotard, and numerous other luminaries—could still possibly cling to self-evidence as an intrinsic property of alleged quasi-objects called facts, propositions, and logical principles? Have we not learned that nothing is evident in itself (i.e., self-evident), and that what has been called "self-evident" is nothing but alleged facts and propositions that we do not see any need to challenge for the present moment?

Self-evidence is not an intrinsic property of anything, because (1) it is not a property, and (2) if it were, it could not be intrinsic. To rehash familiar arguments, let us begin with the observation that evidence is a matter of what types of phenomena are allowed to count as relevant in support of epistemic claims made within particular communities of inquirers. Any such community of inquirers will have a history characterized by specific interests and goals of inquiry, by the system of practices (in MacIntyre's sense) that constitute the tradition, and by the frequently changing technologies of inquiry and production that partially define a particular tradition. What is allowed to count as evidence within such a given framework of inquiry will depend on which phenomena we take to be most important relative to our experimental techniques, our values, and our goals of inquiry. A particular claim would be self-evident *only* in the sense that, at *this* particular moment in time, for *this* particular community of inquirers, with *their* particular set of values and goals, and relative to *these* currently accepted methods of inquiry and modes of explanation, then we members of the community, *for the present*, have no reason to challenge what is posited as "self-evident." We simply proceed, in that particular situation, as if the claim were true. However, as the history and philosophy of science in the twentieth century have shown, virtually nothing is immune to criticism and possible revision or rejection, given changing conditions of the situation (Quine 1960; Kuhn 1970; Rorty 1979, 1982).

If there is any plausible notion of self-evidence (which I seriously doubt), it can only be in the very limited sense recognized by Dewey, who says that to say of a logical proposition that it is self-evident "means that one who reflects upon it *in the meaning system of which it is a member* will apprehend its meaning in that relation—exactly as one might apprehend the meaning, say, of the empirical proposition 'that ribbon is blue.' The question of the logical force and function of the proposition, of the interpretation to be given it, remains open—just as does the *truth* of the empirical proposition after its *meaning* is grasped" (1938/1991, 158; italics in original). Dewey is saying that the only thing that could legitimately be called "self-evident" is the *meaning* of a proposition within a conceptual scheme, but *never its truth!* In other words, it might be acceptable *within a conceptual or meaning system* to say that a proposition's *meaning* is clear and unquestionable (= self-evident), but that gives us no basis guaranteeing the *truth* of the proposition, since that can only be determined relative to standards accepted within epistemic communities. Schafer-Landau is mistakenly equating clearly understanding the meaning of a proposition within a particular context with grasping it as a foundational truth.

In short, for the notion of self-evidence to do any significant moral work, it would depend on (1) the conceptual system or system of symbolic meaning employed (Putnam 1981), (2) a communal determination of what is allowed to count as evidence or data, (3) the community of inquirers having established methodological criteria for explanation and justification, and (4) the assumption of certain values as goals of inquiry. This is precisely where moral realism goes wrong, insofar as it follows Schafer-Landau's claim that some facts obtain and some propositions are self-evidently true, completely independent of any observer, community of observers, or participants in an activity.

There Are No Moral Facts Worth Having

So far, I have challenged the notion of self-evident truths. I will now challenge the very notion of a *moral fact*, whether self-evident or not. I will argue that (1) there is no notion of a moral fact capable of doing the foundational epistemic work of justification that the term "moral fact" was designed by moral realists to accomplish, and (2) to the extent that there is any reasonable sense of the term "moral fact," it is compatible with the non-foundationalist, inquiry-oriented theory that I am proposing.

Let consider more closely the notion of a fact. Moral realism requires a conception of moral facts as foundational givens upon which further knowledge can be erected. We may have to discover facts by long and difficult intellectual labor, but, according to the moral realist, we do not make or create the facts. The "facts" are "out there" waiting for us to encounter or discover them. They are objective givens. As we saw earlier, according to Schafer-Landau's realism, it is an unquestionable fact that "it is wrong to take pleasure in another's pain, to taunt and threaten the vulnerable, to prosecute and punish those known to be innocent, and to sell another's secrets solely for personal gain" (2003, 248). Apparently, there would be four moral facts here, each coming with the label "self-evident" epistemically stamped on it for any inquirer who "correctly" and "fully" understood the meanings of the terms used to express the facts.

I do not wish to challenge any of these alleged principles. Indeed, I would expect to find principles of this sort in most, if not all, moral systems. Strong arguments for principles like these could no doubt be framed relative to many different cultural institutions and moral practices. However, none of this lends any credence to the moral realist notion of a moral fact.

I want to suggest an alternative Deweyan conception of the nature of a fact that in no way supports any foundationalist pretensions of the sort adopted by moral realism. As we saw earlier in his *Logic: The Theory of Inquiry* (1938), Dewey situates logic within practices of human inquiry. Logic is not an autonomous structure of a transcendent, universal rational capacity that is somehow brought to bear to constrain our reasoning. Rather, logic involves the patterns of inquiry that we (a certain community of inquirers) have found useful for pursuing our various modes of engaged, embodied, embedded investigation of the cosmos. According to this view, facts are *posited or proposed* as a way of constraining and carrying forward a certain specific inquiry. They are not preexistent absolute givens to which any correct theory must somehow conform.

Dewey's important proposal here is that we should regard "facts" and "ideas" as *functional* notions arising within a process of inquiry, rather than being abstract quasi-objects. To posit some state of affairs as a "fact" is to circumscribe a problematic situation as having a specific character, which then gives rise to "ideas" about how the situation will develop under certain observed conditions.

> It was stated that the observed facts of the case and the ideational contents expressed in ideas are related to each other, as, respectively, a clarification of the problem involved and the proposal of some possible solution; that they are, accordingly, functional divisions in the work of inquiry. . . . Ideas are operational in that they instigate and direct further operations of observation: they are proposals and plans for acting upon existing conditions to bring new facts to light and to organize all the selected facts into a coherent whole. . . . What is meant by calling facts operational? Upon the negative side what is meant is that they are not self-sufficient and complete in themselves. They are selected and described, as we have seen, for a purpose, namely statement of the problem involved in such a way that its material both indicates a meaning relevant to resolution of the difficulty and serves to test its worth and validity. In regulated inquiry facts are selected and arranged with the express intent of fulfilling this office. They are

not merely *results* of operations of observation . . . Being functional, they are necessarily operational. Their function is to serve as evidence and their evidential quality is judged on the basis of their capacity to form an ordered whole in response to operations prescribed by the ideas they occasion and support. (Dewey 1938/1991, 116–17)

A fact is thus a selection of certain aspects of a situation as a means of ordering your inquiry into that situation. Once you temporarily settle on the "facts" of a situation, you can then employ ideas about the meaning of the situation as a basis for trying to solve the problem as it is thus defined. Dewey therefore concludes that "the order of facts, which present themselves in consequence of the experimental observations the ideas call out and direct, are *trial* facts. They are provisional. They are 'facts' if they are observed by sound organs and techniques. But they are not on that account the *facts of the case*. They are tested or 'proved' with respect to their evidential function just as much as ideas (hypotheses) are tested with reference to their power to exercise the function of resolution" (117).

The upshot of this functional and experimental conception of facts as provisional evidential posits is that *there is no place for facts as self-evident truths* of the sort required for Schafer-Landau's moral realism. The test of a process of moral deliberation is not whether it conforms to some allegedly pre-given foundational facts, but instead whether and how well it resolves a problematic situation. Since in any serious moral inquiry a crucial stage is defining the problem at hand, there will be a critical role for the provisional proposing of certain facts, just insofar as their role is to characterize the problem as you experience it. However, your subsequent ideas about possible ways to resolve the problematic situation might sometimes turn out to lead you back to a redescription of the problem—a reconsideration of the "facts" on which you based your moral deliberation.

This Deweyan conception of a "fact" as non-foundational in any absolute sense does not support any interesting or usable notion of epistemic self-evidence. It is thus perfectly compatible with the account of moral cognition as a non-foundational, process-oriented mode of situated moral problem-solving that I have been developing.

A Non-Foundationalist Conception of Moral Principles

Just as the positing of certain facts is part of a process of moral inquiry, so also is the consideration of moral principles. They are both elements to be factored into one's inquiry into how best to resolve a morally problematic situation. Principles are experimental hypotheses to be tested by actual moral inquiry, to determine if and how they help us find a good resolution of a moral problem. Principles are summaries of the collective wisdom of a people, insofar as they indicate considerations that a moral community has found important for guiding their problem-solving in certain kinds of cases in the past. They are not absolute, pre-given rules for how to judge a particular case. They have a certain presumptive weight in our moral deliberations—because they have a measure of cognitive entrenchment—but they are not absolutes. They arose in the context of certain historical conditions, and they may need to be revised in light of changed, hitherto unanticipated conditions. Dewey defines principles as

empirical generalizations from the ways in which previous judgments of conduct have practically worked out. When this fact is apparent, these generalizations will be seen to be not fixed rules for deciding doubtful cases, but instrumentalities for their investigation, methods by which the net value of past experience is rendered available for present scrutiny of new perplexities. Then it will also follow that they are hypotheses to be tested and revised by their further working. (1922, 165–66)

From this perspective, consider how we should regard Schafer-Landau's allegedly self-evident principle that "it is wrong . . . to prosecute and punish those known to be innocent." This may well be a principle that no American raised in a Western democratic society of the sort operative in the United States today (and having a long history) would ever question. I would expect that any well-educated and moderately reflective American citizen might be able to give you several good reasons why we ought to abide by this principle. Most of these justifications would center, I would guess, on matters of what it means to have a just society and what basic freedoms are requisite for the ideal democracy most of us could envision. None of this justification, however, would establish this as an absolute, self-evident moral proposition. Nor do we need any such absolute justification. All we have ever had is whatever arguments for abiding by such a principle that we can summon up within the justificatory practices of our society as it is currently constituted. We do not need, nor are we able to provide, any proof that such a principle came from God, or universal pure practical reason, or any other transcendent source.

The only way a principle could be fixed, complete, absolute, and final is if our world was immune to change—frozen in its essential patterns. Change, however, is basic to the very possibility of experience, which is inescapably temporal and undergoing transitions. The changing, temporal character of experience thus requires flexibility of thought, adaptation to changed conditions, and reconsideration of fixed habits, conceptual systems, and sometimes even moral principles. Moral principles are never to be taken lightly, but neither are they to be taken as absolutes. They are guides to important considerations that ought to be tried out and tested in our ongoing moral deliberations.

In concluding my challenge to moral fundamentalism from the perspective of cognitive science, I should point out that my arguments apply equally to classical deontological and consequentialist approaches. I have focused primarily on moral law theories, which tend to take the form either of divine commands or rational imperatives. It is clear that most versions of utilitarianism are subject to the same criticisms. Everything I have said so far applies to any form of utilitarianism that is predicated on the following assumptions: (1) that action is right that maximizes the good for the largest number of relevant subjects; (2) the concept of the good is fixed, pre-given, and well defined; and (3) the most efficient means to the realization of the recognized good are calculable. Scores of philosophers and psychologists have definitively criticized these and other grounding assumptions of classical utilitarian theory, and I will not rehearse any of those arguments here. I simply want to note that utilitarianism so defined is just one more version of moral fundamentalism and, as such, is undermined by challenges to fixed, pre-given goods and purposes, and also to criticisms of algorithmic forms of reasoning that are utilized in rational actor models so popular in utilitarian theories. However, let me be clear in acknowledging that none of the challenges I have raised in this chapter need be applicable to a suitably naturalized form of what Kitcher calls "dynamic consequentialism," which he summarizes as follows:

The ethical project can be understood as a series of ventures in *dynamic consequentialism*. Participants in it respond to their problems by trying to produce a better world. The have an implicit conception of the good and take rightness of actions to depend on their promotion of the good as they envisage it. . . . [B]ut a consequentialist ethical theory can explicitly acknowledge it has no complete specification of the good, seeing its judgments as incomplete and provisional. (2011, 288; italics in original)

Dynamic consequentialism is an appropriate name for the type of nonabsolutist view in which we are concerned with the large-scale consequences and states of affairs realized by various choices, practices, and institutions, and in which our perceived ends and notions of the good life are necessarily works-in-progress.

CHAPTER SEVEN

Why Moral Fundamentalism Is Immoral

At this point, we have undermined the alleged foundations of a strong version of moral realism, which, you will recall, was formulated to insure an objectivity that was supposedly entirely independent of any individual or group perspective. There is no perspectival-less objectivity. There are no absolute, self-evidently true moral facts, nor are there any absolute and unconditional moral principles. There is no adequate cognitive or epistemic foundation for moral fundamentalism.

Moral fundamentalism is thus a false doctrine because it is incompatible with what we have learned about how our minds work. Humans do not, for the most part, have classically structured moral categories of the sort required for moral absolutism. Neither do we have strictly literal moral concepts that could map directly and determinately onto fixed states of affairs in the world. Neither facts nor principles can be regarded as "self-evident" in any significant sense of that term, as required by strong moral realism (as a form of fundamentalism). This is why I have called moral fundamentalism a "non-starter" from the perspective of the cognitive sciences, neurosciences, and pragmatist conceptions of mind, thought, language, and values.

But moral fundamentalism is not just *false* from a scientific perspective. Even worse, it is *immoral*, and we can now see why. What makes it immoral is that it shuts down any serious form of moral inquiry, since it takes moral truths as self-evident givens. It denies the complexity, depth, richness, and changing character of experience, insofar as it selects one basic consideration as the sole key to the determination of a morally problematic situation. Even worse, it shuts out any further moral inquiry, on the assumption that experience forms a closed, fixed system subject to timeless laws. Whistling "it was good enough for granny, so it's good enough for me" is immoral. What if granny was a racist, an anti-Semite, a pedophile, a warmonger, a sadist? Neither the faith nor the morality of our "fathers" (or our forefathers) should be regarded as "good enough for us," at least in any absolute sense.

What we need are not unconditional moral truths that never change, for there are no such things available to human beings in any absolute or eternal sense. What we need, instead, are refined, tested, and sensitive methods of moral inquiry that we bring to bear on our lives, testing them by our ongoing experience and adjusting them by our best lights, in the face of changing conditions. As Dewey expressed it:

Morals must be a growing science if it is to be a science at all, not merely because all truth has not yet been appropriated by the mind of man, but because life is a moving affair in which old moral truth ceases to apply. Principles are methods of inquiry and forecast which require verification by the event; and the time honored effort to assimilate morals to mathematics is only a way of bolstering up an old dogmatic authority, or putting a new one upon the throne of the old. (1922, 164)

The search for absolute foundations is a fool's quest. It leads us away from the very thing we most need if we are to face the moral problems that daily confront us—namely, an intelligent, reflective, self-critical, and imaginative moral inquiry that leads us to reasonable action. This will be an ongoing transformative process, so long as we live.

EIGHT

The Making of a Moral Self

What Should a Moral Theory Do?

Here, in outline, is my summary of the argument I have been developing so far for an empirically responsible account of a morality that is fit for humans as we currently know them.

1. Human moral agents are highly complex animals whose values emerge in ongoing interactions with their physical, interpersonal, and cultural environments. Those values arise naturally and are not handed down from some transcendent source.
2. Our values relating to human well-being come from many sources: the biological necessities for life maintenance, aspects of our interpersonal relations, requirements for group cooperation, and considerations of individual and group flourishing.
3. Experience does not come pre-classified into distinct types, so there is no such thing as an exclusively moral experience. Therefore, there is no need to posit any scientifically suspect moral instinct, moral organ, or discredited faculty psychology to account for any allegedly distinctive moral judgments.
4. Instead of thinking of moral reasoning as the application of uniquely moral rules to concrete situations, we should see ethical thinking as nothing but a type of problem-solving activity. All forms of knowledge, including the results of scientific research, can contribute to the intelligence, quality, and effectiveness of our moral problem-solving by informing us about who we are, how we evolved to our present condition, how we think and feel, where our values come from, and what kinds of institutions and practices have existed throughout our history. Consequently, every type of empirical

knowledge and every exercise in human meaning is pertinent to deciding how we ought to live. This ranges over history, biology, neuroscience, anthropology, psychology, sociology, religion, the arts, and the humanities.

5. The idea of moral absolutes, whether in the form of universal moral laws or foundational moral facts, is challenged by cognitive science research into the nature of human conceptualization, understanding, appraisal, and reasoning. Moral fundamentalism, besides being *scientifically suspect*, is also *morally suspect*, insofar as it silences the ongoing moral inquiry we need most if we are to have any hope of dealing intelligently with our problems. There can be a role for moral ideals or principles, since they are summaries of values, considerations, and strategies that past experience has shown to be useful in resolving certain types of moral problems. However, given that the conditions of human existence are both complex and frequently changing, we cannot always rely on existing standards that may have arisen in situations substantially different from those we currently face.
6. The sciences of mind have recently investigated two fundamental processes of moral cognition: (a) a mostly nonconscious, non-reflective, fast, and affect-based intuitive process of appraisal, and (b) a conscious, reflective, slow, after-the-fact justificatory form of reasoning that tends to be principle-based.
7. In addition to these two processes, I have argued that there is a third process—imaginative moral deliberation—in which we imaginatively simulate (i.e., rehearse) possible courses of action, in order to determine which course best resolves at least some of the tensions among competing values, interests, and ends that we are currently encountering.
8. This third form of appraisal involves feelings and emotions (of the sort operative in the intuitive track), but it also has a more reflective, critical, and exploratory process that can properly be called a form of imaginative reasoning.
9. Consequently, moral deliberation is principally a process of assessment of the adequacy of various imagined courses of action for reconstructing our problematic situation into a more fluid, harmonious, and fulfilled reality. A deliberative process is reasonable when it actually gives us a sense of resolution of conflicting ends and values. Reasonableness is thus an outcome of imaginative deliberation, not the aligning of our deliberation with some allegedly pre-given standards of rationality.

The Metaphysics of Moral Experience: Three Metaphorical Conceptions of the Nature of Moral Thinking

The account I have just summarized requires a view of the nature and purpose of a moral theory that is substantially different from the dominant conception in most Western philosophical traditions. The change of view that is necessary cannot be achieved by minor tinkering with received moral theories; instead, it demands a basic overhaul of some of the foundations of those theories. In particular, it requires us to abandon the mistaken idea that moral philosophy ought to provide absolute standards of governance for human behavior.

What is at stake here is not just our conception of moral theory, but our entire metaphysics—our ontological picture of the world. The chief question amounts to this: Is reality a closed system of preestablished essences of things, persons, and events manifesting their intrinsic natures in the world? Or, as I have argued, is experience a process of ongoing interactions of organisms with their environments, marked by change and the emergence of new conditions? The former is an objectivist metaphysics, and the latter is a pragmatist process metaphysics.

The underlying objectivist metaphor sees moral reasoning as an act of *discovery* of something (*viz*., the moral standards) that exist outside and independent of the situations to which they are to be applied. Within this metaphorical framework, experience supposedly comes fixed and pre-categorized into action types, the moral rules or standards are essentially fixed, and the only form of reasoning is recognizing individual cases as fitting pre-given moral types or categories that are then subsumable under highly determinate moral rules. There can be no serious addressing of the newly emerging conditions of a situation that might require solutions hitherto unappreciated and untried.

The idea that there are such foundational moral truths discoverable by revelation, by reason, or by feeling is not, I have argued, supportable by research on how humans conceptualize, value, and reason. Rejection of the *discovery* metaphor has, unfortunately, often pushed us toward the opposite extreme of adopting a metaphor of *arbitrarily making* or *constructing* moral values and laws, rather than *finding* them.[1] From the perspective of the cognitive sciences, this is an overly simplistic and quite unsatisfactory either/or logic that offers us only two radically opposed and equally mistaken descriptions of the process of moral judgment; namely, either (1) as the *discovery* of external or transcendent constraints on morality, or else (2) as the arbitrary *creation* of values *ex nihilo*. In other words, either we *find* those moral values objectively existing in the world, or else we just *make them up!*

In sharp contrast, the pragmatist view adopts a process- and event-oriented metaphysics of experience that recognizes *both* change *and* relative stability in our lives. A world with this sort of character requires a different metaphor, one of *creative transformation* of our experience—a form of artistic remaking of experience. It is a metaphor that stresses the *art* of living rightly and well, via a form of constrained yet creative problem-solving activity more akin to the making and judging of artworks than to the discovery of preexistent truths. In avoiding the dichotomy between *values as discovered* and *values as made*, we must not slip into the relativism of denying that there are any constraints on, or grounds for, our values. I hope it is clear from chapter 2 that we do not merely "make up" our values, just as it is clear from chapters 3–5 that our imaginative deliberation operates under constraints, and therefore not arbitrarily. The notion that the absence of moral absolutes catapults us into a barren moral wasteland, where anything and everything is permitted, is simplistic nonsense.

According to the pragmatist view I am articulating, when we encounter a morally problematic situation, the initial processes by which we evaluate our situation are mostly nonconscious, operating at the intuitive-track level where our emotional assessment and responses have already begun to selectively narrow the range of options we will consciously consider. On the basis of that earlier affective directing and selecting, we then imaginatively explore the meaning and implications of the range of possible courses of actions that appear to us. To the degree that our moral reflection rises to the level of conscious inquiry, our best strategy is typically to make use of all of the relevant knowledge we can muster from every conceivable discipline, in order to determine where different imagined paths of experience might lead us, and what their broader effects might be.

This appeal to broad and deep knowledge of human affairs and the world is no more than a common desideratum for *any* sort of good problem-solving, of which so-called "moral" problem-solving is but one instance. To ensure that you actually make the situation better through your inquiry, you need to know about how things work in the natural and cultural worlds you inhabit. Relevant knowledge can come from any of a number of modes of inquiry, such as biology, neuroscience, anthropology, psychology, sociology, political science, economics, history, the arts, literature, philosophy, and any other discipline that gives us insight into our world, human nature, psychosocial developmental, and culture. Ongoing scientific research can thus contribute to our moral understanding in at least the following ways:

- It can show us where our values come from and why creatures like us have the kinds of values we do.
- Cross-cultural and historical studies can help us understand how our moral views and theories have emerged within culturally specific, historically contingent situations, in response to particular scientific, political, religious, philosophical, economic,

social, and other conditions and developments. Our values are not, therefore, absolute or history transcending in any radical sense.
- The cognitive sciences can help us understand the working of human conceptual systems, such as how concepts are structured, how abstract thinking traffics mostly in conceptual metaphor, and how our reasoning is grounded in our bodily (sensory-motor) experience and is bound up with our emotional responses and feelings.
- Empirical research can give us insight into the processes of human appraisal, deliberation, and judgment. This can help us avoid cognitively inadequate assumptions about mind, thought, and values that often creep into both common moral understanding and sophisticated moral philosophies alike. The cognitive sciences, in dialogue with philosophy, are our principal source of reliable knowledge about human conceptualization, reasoning, emotional appraisal, and reflective judgment.
- Psychology, anthropology, and sociology can reveal which notions of well-being and flourishing arise in which cultures, and they can explore how these various conceptions are either challenged or supported by our current scientific understandings of bodily functioning, human development, and social cooperation.
- The social sciences can give us insight into which practices, social arrangements, and modes of living contribute to different conceptions of happiness or flourishing.

Ethical naturalism, so conceived, can provide us a sound basis for intelligent moral inquiry. As I have argued, ethical naturalism of this sort is non-absolutist, non-relativist, non-reductive, and ameliorative. It is *non-absolutist* because it understands that there are no non-perspectival, ahistorical, or epistemically pure (or *a priori*) transcendent sources of moral values or principles. It is *non-relativist* insofar as it believes that we can provide evidence and general reasons for thinking that certain possible courses of action are better, and others worse, when it comes to our moral problem-solving within a particular situation. It is *non-reductive* because it realizes that no one method, or no single type of explanatory framework, is by itself adequate to the complexity of experienced situations. It is *ameliorative* insofar as it is predicated on the psychologically realistic hope that collective human reflection and agency can make things better through intelligent problem-solving, just as it can make things worse when it fails to adhere to the highest standards of moral inquiry.[2]

What Kind of Self Ought We to Cultivate?

Insofar as it issues from our habits of appraising, thinking, and acting, our moral inquiry is a manifestation of our character and personality, expressing who we are as presently developed. When we act in routine, habitually engrained ways, there is little or no actual deliberation or reflection involved. As we have seen, this is what Dewey called "intuitive valuing" (Dewey and Tufts 1932/1989, 266; Dewey 1939/1991)—we just value what we have come to value, either as a result of our biological makeup, our acquired personal habits of thought and action, or our learned cultural values and practices, and then we act accordingly.

Genuine moral deliberation, in sharp contrast, arises only when we stop doing what we were mindlessly doing and begin to inquire into our situation that has in some way become conflicted or disharmonious. Only when we reflectively engage a problematic situation, do we have an opportunity to remake ourselves (and our world) by changing our interactions with our environment. On those occasions when we are able to be more reflective about what we ought to do in a given situation (i.e., cases in which we are engaged in Deweyan *valuation*), we have the opportunity to reconfigure our habitual selves in a way that transforms who we are. As Dewey explains, when we recognize that we are confronted with a tension among competing claims, values, ends-in-view, and possible courses of action, we have to go beyond the habits of experience and thought that brought us into the conflicted situation in the first place, and this deliberative going beyond is a reformation of self:

We have to make up our minds, when we want two conflicting things, which of them we *really* want. That is a choice. We prefer spontaneously, we choose deliberately, knowingly.

Now every such choice sustains a double relation to the self. It reveals the existing self and it forms the future self. . . . Deliberation has an important function in this process, because each different possibility as it is presented to the imagination appeals to a different element in the constitution of the self, thus giving all sides of character a chance to play their part in the final choice. The resulting choice also shapes the self, making it, in some degree, a new self. (Dewey and Tufts 1932/1989, 286–87)

The "new" self that arises from our moral problem-solving is, obviously, not an entirely new self fabricated from whole cloth. It is an emerging self-in-process that retains a large measure of continuity with our self as previously constituted, and this includes our prior habits of feeling, imagining, and thinking. Ideally, it is a self that grows and expands, through its appreciation of the new complexities encountered in its current situation, instead of stagnating and atrophying, on the one hand, or fragmenting and disintegrating, on the other. Dewey nicely sums up this self-in-process:

> Except as the outcome of arrested development, there is no such thing as a fixed, ready-made, finished self. Every living self causes acts and is itself caused in return by what it does. All voluntary action is a remaking of self, since it creates new desires, instigates to new modes of endeavor, brings to light new conditions which institute new ends. Our personal identity is found in the thread of continuous development which binds together these changes. In the strictest sense, it is impossible for the self to stand still; it is becoming, and becoming for the better or the worse. It is in the *quality* of becoming that virtue resides. We set up this and that end to be reached, but *the* end is growth itself. (Dewey and Tufts 1932/1989, 306)

The truly radical idea here is not just that the self is a self-in-process, but rather the idea that the end of our reflective transformation of experience should be "growth itself." The moral fundamentalist will object that "growth" alone cannot be the end, because we can always ask, "Why growth? What is growth good *for*?" And that question, they will insist, requires some other additional value that growth would need to be measured by.

By now, it should be clear that the question "What ultimate purpose does growth serve?" is both question-begging and seriously misleading, insofar as it presupposes the very objectivist metaphysics (in the form of moral absolutism) that we have rejected. It erroneously assumes that growth itself cannot be a sufficient value and could only be validated by some other value or purpose that it serves.

Dewey's alternative view is that growth is the end-in-view of good moral inquiry. Because there are no final causes (as overarching human ends), absolute values, or supreme moral laws, what we should strive for is growth of meaning. In other words, once we abandon the objectivist metaphysics of moral fundamentalism—according to which growth would be justified as a value only if it served some other (absolute) fixed and final purpose or value—we would come to recognize growth itself as integral to moral well-being and as the condition for the only kind of moral living that we have any measure of control over as reflective beings. The only way to keep moral fundamentalism from creeping back in is to realize that growth itself is the goal of moral deliberation and the sole test of any process of imaginative reasoning about better and worse.

What Does Growth Consist in, as a Basis for Morality?

The big question, then, is the following: What does growth mean, from the perspective of ourselves as moral-agents-in-process? In order to get an adequate definition of moral growth, we have to remember our prior account of what it is in a given situation that first gives rise to the need for moral reflection. Recall, one last time, that the need for moral deliberation arises only when our established habits of thought, feeling, and action (which were sedimented in response to some prior history) run up against new conditions that they are not fitted for and cannot adequately resolve. Habits and feelings conditioned by earlier situations have become inadequate for dealing with new, changed conditions, and that is why we are experiencing frustration, conflict, indeterminacy, and disharmony. Typically, the situation proves to be more complex than we were prepared for, and the result is that we are unsure what the better course of action might be, because our prior habits (our established ways of feeling and responding) leave us without sufficient resources for addressing the problem(s) we are facing.

What we need is a deeper and broader comprehension of the complexity of the situation in which we find ourselves—a depth of understanding that is adequate to all of the impulses, emotions, ends-in-view, and interests active within our current situation. At least some of these aspects of our situation are in conflict or tension, and so our prior habits of feeling, thinking, and behaving are in conflict. If we are to move forward, we need an enriched understanding of all of the factors operating within our situation and all of the possibilities open to us for harmonizing our competing ends, principles, and values. In a very real sense we need to go beyond our present self-identity:

> The growing, enlarging, liberated self . . . goes forth to meet new demands and occasions, and readapts and remakes itself in the process. It welcomes untried situations. The necessity for choice between the interests of the old and of the forming, moving, self is recurrent. . . . For everywhere there is an opportunity and a need to go beyond what one has been, beyond "himself," if the self is identified with the body of desires, affections, and habits which has been potent in the past. Indeed, we may say that the good person is precisely the one who is most conscious of the alternative, and is the most concerned to find openings for the newly forming or growing self; since no matter how "good" he has been, he becomes "bad" (even though acting upon a relatively high plane of attainment) as soon as he fails to respond to the demand for growth. Any other basis for judging the moral status of the self is conventional. In reality, direction of movement, not the plane of attainment and rest, determines moral quality. (Dewey and Tufts 1932/1989, 307)

The crucial importance of the idea of direction of movement as the key determinant of moral quality has not been fully appreciated in moral philosophy.[3] To repeat the argument: If there is no source of value residing outside experience (and then brought to bear upon it), you are left not with arbitrary creation of values, but rather with growth in depth, richness, and scope of meaning as your only guide. The *last* thing you would want in such a situation is a rigid, unreflective reaffirmation that your moral problem could be solved simply by getting clear which pre-existing determinate standard or principle applies to the case at hand. The *last* thing you need is an unthinking insistence that the right rule of behavior is waiting there to be found, if only we were insightful enough to discover it. The *last* thing you need, in short, is moral fundamentalism. That would be a recipe for continued moral obtuseness, close-mindedness, and refusal to engage in moral inquiry. To the extent that experience is actually changing (which is precisely the reason you are faced with a moral problem in the first place), what you need is a process of moral inquiry that allows you to explore all of the relevant dimensions of your situation. In other words, you need growth of meaning.

The proper "direction" of this process of inquiry is thus increase in the depth, breadth, and complexity of meaning of the situation. The meaning of any object or event is the experiences it evokes.[4] Those experiences can be past (consisting of relations to prior events), present (referring to experiences currently had), or future (involving possible experiences that might follow). If the reason you are encountering a moral problem in the first place is that there are more conflicting components of your current situation than you have heretofore appreciated, then the crux of moral inquiry must be, ideally, to understand all of the relevant dimensions that constitute your present conflicted experience. This is your best hope for intelligent resolution of moral problems. Dewey describes the type of progress involved in such a movement of thought:

The present is complex, containing within itself a multitude of habits and impulses. It is enduring, a course of action, a process including memory, observation and foresight, a pressure forward, a glance backward and a look outward. It is of *moral* moment because it marks a transition in the direction of breadth and clarity of action or in that of triviality and confusion. Progress is present reconstruction adding fullness and distinctness of meaning, and retrogression is a present slipping away of significance, determinations, grasp. (1922, 195)

Growth means the liberation of experience and thought by means of recognizing the complexity of relations within a given situation and imaginatively following out such possible relations in projected courses of action, in order to see if we can harmonize competing impulses, values, and ends, and thereby resolve some of the tension within our conflicted situation.

Why might so many people find this conception of moral growth so objectionable, or at least uninspiring? The answer, I have been suggesting, is that they are laboring under the illusory expectation that moral reasoning *could* and *ought to* discover the one correct principle (or set of principles) that is properly applicable to our present state of affairs, once we have correctly categorized or described it. The appropriate alternative to a pre-given standard or outcome of reasoning is attention to the character of the process of reasoning itself. The more you understand the various meanings, values, emotions, and purposes that make up your present situation, the better you are able to choose an appropriate course of action from among all those available to you from your deliberations. The sin of moral obtuseness (or failed perceptiveness) is the inability to appreciate all, or at least most, of what is going on in the situation you are facing. It is a failure to recognize a plurality of competing values, conflicting habits, incompatible desires, emotional responses, and motivational dispositions. It is also a failure of vision regarding the possibilities for reconstructive improvement of the present situation. The result of such a failure is dogmatic adherence to prior habits of feeling, thinking, and response that only reinforce what is problematic about the situation in the first place.

There can be no other adequate standard, then, than progress in the growth of the self and meaning. Dewey's classic statement of this conception of growth is the following:

Morals means growth of conduct in meaning: at least it means that kind of expansion in meaning which is consequent upon observations of the conditions and outcome of conduct.... In the largest sense of the word, morals is education. It is learning the meaning of what we are about and employing that meaning in action. The good, satisfaction, "end," or growth of present action in shades and scope of meaning is the only good within our control, and the only one, accordingly, for which responsibility exists. (1922, 194)

We are better off, Dewey is saying, the more perceptively, sensitively, and comprehensively we are aware of the distinctions, relations, emotions, and values operating within our present situation. The key to whatever freedom we are capable of is knowing the meaning of our situation so that we can effect change for the better where, when, and to what extent that is possible. The better we grasp the principal relevant factors operating in our present situation, the more intelligently and wisely we have the opportunity to address any problems that arise. We want to harmonize competing values and ends; otherwise, our actions are disjointed, at odds with themselves and the ends of other people, and therefore liable to engender even more difficulties. We want to unify our situation, which means resolving, as far as is humanly possible *right now*, whatever is at odds within and without us. Unification of experience is an ongoing dynamic ordering (i.e., allostasis) of our situation, rather than being merely some fixed and completed structural unity among parts.

We want to develop and expand our capacities for thinking and acting, because that is what makes intelligent inquiry and experiential transformation possible. Given his rejection of absolute rules, Dewey guardedly suggests that we replace Kant's categorical imperative with the pragmatist categorical imperative: "So act as to increase the meaning of present experience" (196). The new imperative is appropriate only insofar as we never forget that it is not an absolute moral law, but rather a useful principle of intelligent moral inquiry.

The human condition is such that no "solution" of a problem can ever be final in any absolute sense. We take a course of action, and perhaps thereby realize a certain good, but in realizing that good, there are other alternative goods we do not, and cannot, then realize. Moreover, since we can never be completely sure of the outcome of actions we undertake, nor the situational changes we either intentionally or unwittingly initiate, every chosen good brings with it potentially problematic unanticipated future consequences. For example, the Industrial Revolution brought us many technologies and goods that have enhanced the quality of human life (e.g., improved food production, shelter, safety devices, and medical technologies); however, along with these goods have come unforeseen consequences that have proved to introduce profound new problems (e.g., resource depletion, pollution, tools of modern warfare, alteration of genetic makeup of crops, etc.). Life thus continually presents us with new conditions, new difficulties, new problems:

> Sufficient unto the day is the evil thereof. Sufficient it is to stimulate us to remedial action, to endeavor in order to convert strife into harmony, monotony into variegated scene, and limitation into expansion. The converting is progress, the only progress conceivable or attainable by man. Hence every situation has its own measure and quality of progress, and the need for progress is recurrent, constant. . . . Progress means increase of present meaning, which involves multiplication of sensed distinctions as well as harmony, unification. (Dewey 1922, 195–96)

An Example of Deweyan Moral Deliberation

Since we do not have values and standards as traditionally construed from an absolutist perspective, we are left with moral problem-solving as embodied, situated growth of meaning. From a Deweyan naturalistic perspective, we are engaged in an empirical moral inquiry that draws on all the sources of knowledge of our world and ourselves that we can muster. I want to illustrate this naturalistic, empirically grounded notion of moral inquiry by putting it to work on the issue of gay marriage that has recently become a topic of vehement disagreement in many societies and cultures around the world. This example shows (1) how our understanding of key concepts (here, *marriage*) can change, as the result of changed empirical circumstances, thereby giving rise to novel moral considerations, and (2) how scientific knowledge can be relevant, and even essential, to the investigation of a moral problem. Science is certainly not the whole story, but it is important to realize how much it can contribute to moral deliberation.

In American society and elsewhere, it is only within the past two or three decades that the issue of gay marriage has been thrust into the limelight. Prior to that, it was mostly taken for granted that marriage ought to be a lifelong bond exclusively between a man and a woman. In many Western societies this conception of marriage has been regarded as an indisputable moral fact—a virtually self-evident moral truth either divinely commanded or else somehow inscribed in the "natural order."

How could the morality of gay marriage possibly depend in any way on empirical knowledge of the sort generated by scientific research? Isn't it just a matter of analyzing what the concept *marriage* means, and then immediately recognizing that it does not "contain" or "include" same-sex partners? Isn't it merely a matter of incisive conceptual analysis of the sort that moral theorists are supposed to be masters of? Well, the answer is no—or, at least, that cannot be the whole story.

In order to see this crucial point, we need to stand back and take a broader historical look at what marriage has meant. We begin with the obvious but important point that the need for moral deliberation about gay marriage arises only because some person or group has challenged the received mainstream view. In other words, somebody must have recognized the problematic character of our current situation surrounding this issue. Otherwise, there is no "problem of gay marriage." Some person or group must make a claim that our current arrangements lead to unfairness and injustice for gays and lesbians. It is only this tension of competing claims, values, and interests that gives rise to a moral problem requiring inquiry and justification.

What we need to do, then, is to inquire into the origin of the alleged problem at hand, in order to determine how and why somebody came see any problem at all. We need to figure out why there has arisen a conflict of views, values, principles, ends. Notice that this is already a question susceptible of empirical study, namely, "What is our received conception of marriage and how (under what conditions) did it emerge historically?" Historical study reveals that many cultures have assumed that heterosexual marriage was either God-ordained, or else that it constitutes the "natural" condition for procreation and the raising of a family. Any other arrangement (e.g., pregnancy out of wedlock, gay marriage, polygyny, polyandry) was taken to be either "contrary to the will of God" or else "contrary to nature."

However, from the very start, historical investigation also teaches us that there is nothing universal about this particular conception of marriage! There have been in times past, and are today, cultures that have not shared this notion of monogamous, heterosexual marriage. A history of marriage practices would reveal multiple types of sanctioned relations—polygyny (one husband, several wives), polyandry (one wife, several husbands), and even extremely rare cases of group marriage (multiple females married to multiple males).

The situation becomes even more complicated when we realize that even within a particular conception of marriage (such as heterosexual monogamy), there can be *different* requirements about what the husband owes his wife, how she ought to be treated, what she may expect regarding material goods and personal freedoms, what the range of permissible living arrangements is, and on and on. The combinations and permutations reveal anything but unanimous agreement about the nature of the marriage relation, even within heterosexual monogamy.

The first empirical result, then, is that there is *no transhistorical, universal conception of marriage* that applies to all times and all cultures, from which it follows that our mainstream American conception of marriage is most certainly not a self-evident fact or truth. Instead, it arose under historically contingent circumstances in response to social, cultural, economic, political, and often religious considerations.

It gets worse. The variability of the concept of marriage shows up even within specific moral traditions, such as Hebrew and Christian religious traditions. In the Hebrew scriptures, polygyny is reported as a common occurrence, and wives were regarded as the property of their husbands.[5] However, even within the narrower confines of Christianity, we find no unanimity of thought on the nature of marriage. For instance, marriage is today regarded as a sacrament in Catholicism, but this was not always the case. It was not until the Council of Trent (in several sessions between 1545 and 1563) that marriage was officially identified in canon law as a sacrament, although it had been regarded as a minor sacrament for centuries prior to that time. In contrast, certain Protestant denominations did not, and still do not, recognize it as such.[6]

If we continue to narrow our focus even more, by looking only at Protestant Christianity, where heterosexual marriage is a moral absolute, there have been and remain serious differences of opinion about the precise form that marriage should take. There are religious sects that support polygyny as an acceptable (or even desirable) marital arrangement. But even if we set those "marginal" sects aside, we still find fundamental disagreement among Christians about the rights and duties of husbands and wives. Many conservative Christians today accept St. Paul's controversial twist that husbands are commanded to love their wives, and "wives, submit yourselves unto your own husbands as unto the Lord" (Ephesians 5:22). That this is but one historically contingent requirement must seem obvious to anyone who is not a Christian, but some Christians nevertheless claim that it is an absolute moral and spiritual requirement of a true marriage, as sanctioned by God.

In short, what the most cursory survey of the history of marriage reveals, even when we confine ourselves only to Judeo-Christian traditions, is a broad array of quite different conceptions of marriage. Our currently preferred conception of monogamous heterosexual unity "until death us do part," arose in response to quite particular historical conditions, and then only relatively recently over the past few hundred years. If, as many claim, the precise nature of marriage is God-ordained, then it is a curious fact that so many Christians can disagree about what, precisely, it was that God ordained.

In short, the "essential" nature of marriage appears to vary across sects and subcultures, and throughout the course of history, and we should not delude ourselves into believing that all sane, moral persons would be forced to acknowledge this allegedly "one true meaning" of marriage as heterosexual monogamy. I am claiming that (1) *asserting* that some principle is absolute does not make it so, and (2) if the basis for the alleged absolute character of some conception of marriage is supposed to be that it issues as a commandment from God, then the scripturally revealed will of God should not involve inconsistent and incompatible assertions, which it most assuredly does within the Hebrew and Christian traditions of interpretation of the holy scriptures.

One can easily anticipate the following response by the person of faith: "So what if there are multiple incompatible views of marriage across, and even within, religious and moral traditions. The fact is that all but one of these must simply be mistaken. God, or my implanted human reason, tells me that marriage means life-long heterosexual monogamy, and that's all we need to know." I would simply respond that, while one can certainly hold to such a view, that view cannot be made consistent with *many* cases in the Old and New Testaments, where it is clear that important figures took multiple wives and concubines.

Let us then ask what purpose marriage is thought to serve, under this heterosexual monogamy conception. This received traditional conception of marriage is based on the ideas that (1) monogamous, heterosexual marriage is the moral (and perhaps spiritual) context for sexual union leading to procreation, and (2) marriage is the natural, and moral, context for the raising of a family. Many have simply elevated this historically evolving view of marriage into what they regard as an unconditionally binding, absolute, and unambiguous moral principle. Taken as an absolute standard, it is supposed to follow that men cannot marry men, nor women, women. Furthermore, if marriage is the proper moral and spiritual framework for the conceiving of children, then gays and lesbians are not fit for parenthood with a member of their gender, since they cannot, by themselves, procreate.

Focusing now on the *purpose* of marriage, the next stage of an empirical inquiry into the morality of marriage would involve an examination of whether any of the original conditions that generated the received conception might no longer hold. Perhaps new conditions have arisen that might have a significant bearing on how we should view the nature of marriage. This is an empirical question. Even the briefest inquiry reveals that some of the conditions and assumptions upon which the received view rests no longer hold and have given way to new conditions and constraints. Consider, for example, how certain biblical conceptions of marriage appear to have often gone hand in hand with the absolute prohibition of homosexuality, regarded as an abomination in the sight of God. It was considered "unnatural" (and against God's laws) that a man should lie with a man (Leviticus 18:22) or, by extension, a woman with a woman. Consequently, the idea of a homosexual couple nurturing and raising a family is a non-starter within a cultural understanding that regards homosexuality as immoral.

I also suspect that the belief that a baby could only be conceived through sexual intercourse between a man and a woman (which has actually been the case for most of human pre-history and history) strongly reinforced the belief that the natural (and moral) framework for the raising of a family was within a heterosexual marriage. However, we now know that this is not the case, as we have developed new reproductive technologies, such as in vitro fertilization and surrogacy that allow for the creation of a baby without intercourse. As recently as forty years ago, a woman had to carry her own baby to term (or nearly so) for it to be born and to survive. It took a man, a woman, sexual intercourse, and considerable good luck (or, sometimes, an unfortunate one-night stand) to have a human baby.[7] It was within this very limited knowledge framework that marriage fit our need for procreation and the nurturing of children. Coupled with prohibitions against homosexuality, it was easy to consider gay marriage morally objectionable.

Today conditions have changed quite dramatically, with modern science and new medical technologies altering the range of possibilities for conception, gestation, and birth. We now have available, for a price, in vitro fertilization. We have the possibility of the sperm and egg coming from donors other than the "parents." We have surrogacy, in which another "mother" can carry an implanted fertilized egg to term. We now have newly emerging technologies of genetic manipulation. In other words, the range of conditions for bringing children into the world has changed markedly since the early days of Hebrew and Christian religion, and we have every reason to believe that they will continue to change in the future as technology develops. Sexual intercourse between a man and a woman is no longer a necessity for procreation. This fact alone is enough to call into question that connection of marriage with heterosexuality.

The other dimension of this justification of marriage, namely, the prohibition of homosexuality on the grounds that it is "unnatural" or "against the natural order" has also been subjected to withering attack over the past several years. In addition to changes in medical technology surrounding fertilization and gestation, there have been substantial changes in our scientific understanding of nature and "the natural," especially as this pertains to the question of the alleged "unnaturalness" of homosexuality. The scientifically discredited notion of natural purposiveness (which posited final causes inscribed in the ontological structure of being) can no longer support any of the earlier absolutist claims about the purposes (as Aristotelian final causes) of organisms. It is not that organisms lack purposiveness (of course, they are purposive); but rather, nature has no final causes—no ultimate natural ends toward which individual things or organisms develop. The doctrine of final causes is a holdover originally from Greek immanent teleology and then later from medieval theological frameworks in which every existing thing is the handiwork of a God who designs each entity to manifest or realize some essence and to serve some purpose in the divine plan. Non-theological versions of final causality, such as Aristotle's, regarded natural processes as directed toward fulfilling some purpose or realizing some eternal form.

In radical contrast to this essentialist teleology, explanations in contemporary science no longer sanction reference to final causes as traditionally conceived. Living things do not exist as an expression of some ultimate end or purpose they were created to pursue and to realize. Organisms are purposive, in the sense that they have evolved to preferentially seek certain states, such as satisfaction of nutritional needs, maintenance of homeostasis within their internal milieu, and avoidance of harm. But they are not aspiring, as agents, to realize some ultimate end they were designed for.

The point is not that absent final causes, everything comes to exist randomly, but rather that such purposiveness as we find it is the product of evolutionary processes that have no preestablished, overarching goal or design. What we call "purposes" arise naturally and immanently within the evolutionary development of living organisms. Consequently, if there are no "final causes" in nature, then homosexuality cannot be against nature's purposes, precisely because there are no final natural purposes to be gone against. Moreover, if you find homosexuality arising in nature (both in so-called human nature and also among other animals), then it is as natural as anything could be. Add to this the wariness within contemporary natural and social sciences against the outdated distinction between "the natural" and "the cultural," and then arguments about what is "natural" lose their force as a basis for moral evaluation. Humans are naturally cultural beings. Because it is "natural" for human beings to grow and develop within cultural systems, we cannot, therefore, cleanly and unconditionally separate out what is natural to a thing, as opposed to what is a product of culture—at least not in any strong sense that would be required to rule out certain sexual practices as "unnatural."[8]

The point I am emphasizing here is that empirical scientific research has undermined any form of strong metaphysical teleology. The argument that homosexuality is unnatural cannot stand. The claim that some practice is "against nature" is merely a way of smuggling in a moral preference where there is no argument or any scientific basis for the claim. Therefore, the idea that one reason we cannot have homosexual marriage is that homosexuality is unnatural (and, supposedly immoral) will not hold up under scrutiny.

In short, today we have a much deeper understanding of the contingent circumstances under which our traditional monogamous, heterosexual model of marriage took shape. *We are in a brave and somewhat scary new world that could not have been imagined in the early centuries of Judaism and Christianity when our traditional conceptions of marriage were being developed.* Some of the beliefs and conditions under which earlier moral conceptions were constructed have changed, and so must our conception of marriage.

Finally, other important empirical findings, such as the constantly growing research on human sexuality and family life, ought also to prove relevant to our moral inquiry into the institution of marriage. To cite just one example, there is now social science research concluding that there appears to be no reason to think that gay couples are any less fit to be loving, kind, and nurturing parents than heterosexual couples are.[9] One recent survey of research on this subject concludes: "The research indicates that parenting practices and children's outcomes in families parented by lesbian and gay parents are likely to be at least as favourable as those in families of heterosexual parents, despite the reality that considerable legal discrimination and inequity remain significant challenges for these families" (Short et al. 2007, 4). Gay and lesbian couples can be bad parents, just like heterosexual couples can, but there is nothing inherently inferior about gay- and lesbian-parented families.

What I am suggesting is that empirical (psychological, neuroscience, sociological, anthropological, and historical) inquiry is crucially relevant to our consideration of the nature of marriage. We saw that even the most superficial historical investigation reveals the absence of any universal conception of marriage, and this holds true even within particular religious traditions. We then saw that empirical evidence from various disciplines undermines the traditional claims that (1) only sexual intercourse between a male and female can produce a human baby, (2) that homosexuality is unnatural, and (3) that gays and lesbians cannot be just as good (or just as bad) parents as heterosexual couples. Those who ignore empirical research of this sort about sexuality, gender, child rearing, and human relationships are left with little more than unsubstantiated (and scientifically unsupportable) assertions about the nature of marriage. The preferred alternative to serious scientific inquiry appears to be theological assertion drawing mostly on citation of biblical passages taken to be the literal word of an omniscient, omnipotent, moral God. Those who choose that path will most certainly find no place for empirical investigation or marshaling of relevant scientific evidence. The only type of disputation that remains is reliance on the authority of biblical texts whose interpretation has to be regarded as self-evident, literal, and not subject to disputed interpretation.

Biblical literalism, however, is highly problematic from a cognitive perspective. As we have seen earlier (chapter 7), it assumes a seriously inadequate and mistaken view of human understanding, conceptualization, and reasoning. Furthermore, it is morally suspect insofar as it ends up requiring the faithful to commit to practices that are completely at odds with our contemporary sensibilities about respect for humans. For instance, consider the biblical literalist who cites Genesis 2:21–24 as a biblical justification of heterosexual marriage: "Therefore shall a man leave his father and mother, and shall cleave unto his wife; and they shall be one flesh" (*KJV*). Unfortunately, the defense of moral views by reference to biblical scripture is a notoriously problematic practice, especially when one presumes to take *all* biblical passages (from the historically canonized books determined by certain ecclesiastical bodies to constitute holy scripture) as the literal and absolute word of God. Once you go down that road, you must accept a host of biblical passages condemning certain practices and commanding certain others that even the most devout Jews and Christians could not, in good conscience, accept as moral absolutes. For, if we were to take biblical narratives and commandments as ahistorical, absolute, God-given, literal stories and imperatives, then we would be subject, by requirements of consistency, to adhere to wildly objectionable moral imperatives of the sort that can be easily culled from the pages of books such as Leviticus or Deuteronomy. To cite just four examples (out of scores that can be gleaned from various books of the Bible): (1) "If a man lie with a beast, he shall surely be put to death; and ye shall slay the beast" (Leviticus 20:15, *RSV*); (2) "If a man shall lie with a woman having her sickness, and shall uncover her nakedness; he hath discovered her fountain, and she hath uncovered the fountain of her blood: and both of them shall be cut off from among their people" (Leviticus 20:18); (3) "He that blasphemeth the name of the LORD, he shall surely be put to death, and all the congregation shall certainly stone him" (Leviticus 24:16); and (4) "If a man have a stubborn and rebellious son, which will not obey the voice of his father, or the voice of his mother, and that, when they have chastened him, will not hearken unto them: Then shall his father and his mother lay hold on him, and bring him out unto the elders of his city, and unto the gate of his place; And they shall say unto the elders of his city, This our son is stubborn and rebellious, he will not obey our voice; he is a glutton, and a drunkard. And all the men of his city shall stone him with stones, that he die: so shalt thou put evil away from among you; and all Israel shall hear, and fear" (Deuteronomy 21:18–21).

It is highly instructive to study the pages of requirements set forth in Leviticus for the intricate preparation of burnt offerings as a cleansing ritual for moral misdeeds and for acts that make one unclean ("acts" such as menstruating and giving birth). Cases such as these, and there are scores of them in Leviticus alone, surely reveal the historical, cultural situatedness and contingency of many biblical commandments that any moderately reflective person would recognize as wholly inappropriate for our present age. These are good examples of what most Westerners would today regard as excesses of purity rituals in early religion, based on an erroneous understanding of the nature of menstruation and birth. Prescriptions for the cloistering of a woman having her period, or for the cleansing of her following childbirth, can be no more than historical artifacts of a culture uninformed about the nature of biological processes and obsessed with purification rituals of the sort found in all religions that stress culturally defined notions of spiritual purity.[10]

The point of citing cases like these is to indicate the perils of any kind of literalism and fundamentalism, either biblical or otherwise. No doubt, some people do attempt to pursue a strategy of strict biblical literalism, but the moral costs are extremely high, requiring adherence to beliefs that are demonstrably false, such as the belief that menstruation and childbirth render a woman morally unclean. Another disastrous cost would be the violation of one moral stricture (such as the requirement to love and nurture one's children, or the requirement not to murder) in order to adhere to some other stricture (such as the requirement to deliver up one's rebellious son to the elders so that he can be stoned to death). And while I can imagine that there might be a few fanatics who want to kill a person guilty of bestiality, would they also think they had to visit the same fate on the poor beast that had the misfortune to be chosen by the sinner?[11]

Moral fundamentalism of this sort is designed to end moral debate, not to encourage and enrich it. At the very least, such a posture expresses one's refusal to engage in reflective moral inquiry. At the worst, it denies the fact that our interpretation of any principle, command, or scripture is constrained by hermeneutical practices deemed appropriate within particular communities and subcultures. It denies the historically contingent circumstances in which certain views, values, practices, and institutions arose, and it fails to recognize that changed conditions can, from time to time, require us to rethink our received views.

To summarize: I have been characterizing "growth" in moral deliberation as the bringing into play of deeper, richer, and ever more comprehensive understandings of the situation we find ourselves in. At the heart of this process is, first, an examination of the historical contingencies under which certain values were articulated, certain principles formulated, certain habits developed, and certain interpretive practices taken to be authoritative. We then have to ask whether our current situation is different—and in what ways—from those earlier conditions. In our present case, the "problem of gay marriage" has arisen today under very different conditions than those obtaining during the early Hebrew and Christian eras. The traditional prohibition of gay marriage historically went hand in hand with the prohibition against homosexuality. It also went hand in hand with certain assumptions about marriage being established for the procreation and rearing of children. I am claiming that it took a better scientific understanding of biology and psychosexual development in order to realize that homosexuality is a "natural" condition of certain humans, and not a grotesque deviation from "proper" human nature. Once this is recognized, we are well advised to reevaluate our conception of marriage, as follows.

First, if homosexuality is not unnatural, then perhaps matrimony between two persons of the same sex or gender might be perfectly reasonable.

Second, we needed to realize that procreation does not necessarily require male-female sexual intercourse, so that the having and rearing of children does not necessarily require a heterosexual married couple.

Third, we had to see that parenting is not intrinsically an exclusively heterosexual capacity, since loving nurturance, support, care, and responsibility are capacities of all suitably developed, non-sociopathic human beings.

Fourth, we had to appreciate that it is morally irresponsible to oversimplify by ignoring the complexity of the situations that called us to engage in our moral deliberation in the first place.

Only under something like these conditions can we realize that our received moral principles about marriage might need to be reconsidered in light of newly emerging conditions that did not exist in earlier times. Refusal to examine our changing situation is a form of self-inflicted moral blindness. Moreover, once the need for reconsideration (Dewey's valuation) is recognized, there is no good reason to deny the relevance of empirical considerations arising from the natural sciences, social sciences, and humanities. In sum, there is nothing unusual or inappropriate about the claim that moral deliberation of this sort is an eminently empirical affair that can draw on our best scientific understanding of every aspect of existence, without limiting the sources of insight to science alone. In addition to scientific research, literature and the arts are obviously extremely important sources of moral insight. Short stories, novels, and poems can capture the narrative structure of our lives and our self-identity, in a way that carries us beyond the methodological limitations of the natural and social sciences.[12]

I end this exploration of the role of empirical inquiry in moral deliberation with the suggestion that what we have just seen about the nature of the concept of marriage holds equally for all or most of our moral concepts, however we circumscribe the domain of the moral. *Marriage* is a garden-variety moral notion, representative of the way our moral conceptions operate generally. Like all concepts, it developed within some historical context, relative to the practices, values, and prior conceptual systems of a society. It has a radial category structure (Lakoff 1987), characterized by central (prototypical) cases that shade off into less central members of the category that share some, but not all, of the principal features of the prototype. Therefore, it is not definable by a set of necessary and sufficient conditions for its application. Furthermore, as historical conditions continue to change, so also do some of our concepts change and grow. Our concepts are not known "in themselves," via allegedly fixed structures that persist for all time. They are historically evolving functional notions that can, and ought to be, interrogated and reworked as experience requires.

Developing Conscientiousness

I have argued that the answers, such as they are, to our moral problems are not eternal verities given for all time in advance of our moral deliberations. Moral insight is the result of moral inquiry, not a precondition of it. We have to give up the metaphor of moral truths inscribed on metaphysical, noumenal, or transcendental stone, waiting eternally to be discovered or grasped by us. Moral deliberation is a process of inquiry and not a means to some independently existing truth, value, or state of affairs. We can describe how we ought to deliberate, but we cannot say, in advance, what the outcome of any such deliberation ought to be. That, I submit, is the limit of our moral understanding.

Life is not a game to be played in accordance with rules supplied in a cosmic rule book given to us from beyond our experience. Rather, as Alasdair MacIntyre (1984) has argued, such rules as there are emerge only

CHAPTER EIGHT

within the practices themselves, subject to the historically changing conditions of those evolving practices. Translated into human life and moral traditions, this means that the constraints and guidance we finite, fallible creatures have are those that emerged from the contingent characteristics of ourselves as presently formed and the changing structures of our physical, interpersonal, and cultural environments.

We are thus brought, at last, to the key question of what kind of person-as-moral-inquirer we ought to cultivate. Moral valuation is a boots-on-the-ground operation, to be assessed relative to how well it resolves our problematic situation within our present circumstances in the world. In *Theory of Valuation* (1939), Dewey described the appropriate reflective process of problem-solving as follows:

> There is no a priori standard for determining the value of a proposed solution in concrete cases. A hypothetical possible solution, as an end-in-view, is used as a methodological means to direct further observations and experiments. Either it performs the function of resolution of a problem for the sake of which it is adopted and tried or it does not. (232)

Dewey then proceeds to illustrate this process by analogizing it to best practices with respect to maintaining or restoring health:

> While there is no a priori standard of health with which the actual state of human beings can be compared so as to determine whether they are well or ill, or in what respect they are ill, there have developed, out of past experience, certain criteria which are operatively applicable in new cases as they arise. Ends-in-view are appraised or valued as *good* or *bad* on the ground of their serviceability in the direction of behavior dealing with states of affairs found to be objectionable because of some lack or conflict in them. They are appraised as fit or unfit, proper or improper, *right* or *wrong*, on the ground of the *requiredness* in accomplishing this end. (233)

Moral deliberation is thus goal-directed but not directed to some goal standing outside our present situation, as if it were an ultimate value. The "goal," instead, is problem-solving, or the resolution of an indeterminate situation. The only indication we can ever have that we are perhaps on a good track is that our deliberations issue in reworkings of ourselves and, correlatively, our world, in a way that resolves tensions, conflicts, disharmonies, rigid stasis, and uncontrolled randomness. Our only indication of resolution is our present sense of release from the oppression of our prior situation—right here and now, rather than in some non-existent ideal, future world.

If I have accurately characterized the situations that give rise to moral inquiry, then the kind of moral personhood we should nurture in anticipation of such situations is the opposite of the kind of self that is envisioned in traditional Western moral theories that claim to discover objective, preestablished moral standards! In those traditional theories, the self has an essence that exists prior to, and independent of, the acts it chooses. Such a self is allegedly a knower of moral truths (i.e., truths about values, ends, or principles) that it supposedly brings to bear to clarify what ought to be done in a particular case. According to the Moral Law folk theory that I have been criticizing, the locus of the self is a rational agency that stands under governing principles stemming from the essence of human rationality (or, alternatively, from divine reason and thus discernible by reason). The basic relation between the self and rational principles is constraint. In Kant's words, "All imperatives are expressed by an *ought* and thereby indicate the relation of an objective law of reason to a will that is not necessarily determined by this law because of its subjective constitution," and "consequently, imperatives are only formulas for expressing the relation of objective laws of willing in general to the subjective imperfection of the will of this or that rational being, e.g., the human will" (1785/1983, 413, 414).

According to this objectivist view, the will is given, as a free power to act, prior to its encounter with any morally binding principles of action. Such a self *chooses* the ends it will pursue but is not essentially determined by those ends. There is no new self in the making; rather, there is only an essential self attempting to bring the world into conformity with the preexisting governing values or principles it allegedly has discerned, either through divine revelation, reason, sentiment, or social inculcation. On this static objectivist view, the self is not really *solving* a problem through action in the world. Instead, it only clears up what was supposedly required from a correct understanding of whichever value, end, or principle it properly recognizes as applying to the case at hand. This is the logic of judgment that, as we saw earlier, Donagan (1977) described as discerning what a given moral concept (e.g., the concept of respect for a person) means "in itself."

In radical contrast with this traditional picture of the self, Dewey proposed that the self does not exist independent of its actions but is manifested in and through them, so that ongoing activities are processes of self-reformation. Dewey concludes that "it is not too much to say that the key to a correct theory of morality is recognition of the *essential unity of the self and its acts*, if the latter have any moral significance; while errors in theory arise as soon as the self and acts (and their consequences) are separated from each other, and moral worth is attributed to one more than to the other" (Dewey and Tufts 1932/1989, 288). Dewey's moral agent is a doer and re-former of experience, and not just a knower seeking conformity with guiding principles. An agent of this sort constantly remakes its world, and in the process remakes itself, its character, and its identity.

We need to give up the idea of a preestablished self that seeks to conform its choice to pre-given standards allegedly determined by God, reason, society, or the world. Such a self is precisely what we do not want when we encounter genuine moral problems, for those problems arise only when our prior habits of thought, feeling, valuing, and action are inadequate to the complexity of our new situation. In other words, we have those problems in the first place precisely because our self as presently formed is out of balance with its environment. In such situations, our very self-identity is at stake, because it is a question of what kind of person we are becoming. We cannot simply rest with the preestablished—either in the form of a preset essential self or in the form of pre-given moral rules. To cite a crucial passage once more, what we need, instead, is to transform our problematic situation through reflective inquiry:

> Perhaps the most striking difference between immediate sensitiveness, or "intuition," and "conscientiousness" as reflective interest, is that the former tends to rest upon the plane of achieved goods, while the latter is on the outlook for something *better*. The truly conscientious person not only uses a standard in judging, but is concerned to revise and improve this standard. He realizes that the value resident in acts goes beyond anything which he has already apprehended, and that therefore there must be something inadequate in any standard which has been definitely formulated. He is on the *lookout* for good not already achieved. Only by thoughtfulness does one become sensitive to the far-reaching implications of an act; apart from continual reflection we are at best sensitive only to the value of special and limited ends. (Dewey and Tufts 1932/1989, 273)

The kind of person we should aspire to become is a *conscientious* person. This is the opposite of the dogmatic mind that, convinced of its hold on absolute truth, and suffering under the illusion of a fixed metaphysics of the self and the world, treats moral thinking as rule mongering—the bringing of concrete cases under definite moral principles. Conscientiousness, by contrast, requires the mental and emotional flexibility to imagine new solutions and new ways of going forward that resolve pressing moral problems.

We Are Not Little Gods: Morality for Humans

The central theme of this book is that we need a morality fit for actual human beings, and an important part of that story will come from cognitive science research to help us develop a realistic account of where our moral values come from and how we can address our concrete moral problems through an imaginative process of moral deliberation. At the very end of *Experience and Nature* (1925), Dewey summarizes his key claims by reminding us that we are not "little gods" but rather natural human creatures for whom all experiences, identity, values, judgments, and ideas emerge from our ongoing natural engagement with our world:

> Yet if man is within nature, not a little god outside, and is within as a mode of energy inseparably connected with other modes, interaction is the one unescapable trait of every human concern; thinking, even philosophic thinking, is not exempt. This interaction is subject to partiality because the human factor has bent and bias. But partiality is not obnoxious just because it is partial. . . . What is obnoxious in partiality is due to the illusion that there are states and acts which are not also interactions. Immature and undisciplined mind believes in actions which have their seat and source in a particular and separate being, from which they issue. (1925/1981, 324)

Virtually all of my critical arguments—against moral fundamentalism, the Moral Law folk theory, pure practical reason, and other foundationalist and transcendent perspectives—are based on the inescapable primacy of organism-environment interactions as the basis for our values, meanings, thoughts, and actions. The Moral Law folk theory, of which I have been so critical, starts with a supernatural God as the absolute transcendent source of moral principles and laws, and then, when that is called into question, puts God inside us (in the guise of universal moral reason and conscience), attempting to turn us into "little gods" who are self-legislating absolute sources of moral law.

We are not little gods—we are human animals. Once we get over the illusion that each of us has the divine spark within, we will recognize that we face life with only our animal cunning, our human intelligence, our imagination, and our cultural resources. Once we recognize ourselves as dynamic stabilities realized in processes of making and remaking meaning, self, and world—only then can we begin to develop a self-understanding that is adequate to our embeddedness in intertwined biological and cultural matrices that constitute nature as an ongoing process, or what Dewey calls an "affair of affairs." Any realistic moral understanding has to grow out of our bodily and mental situatedness:

> When a man finds he is not a little god in his active powers and accomplishments, . . . [w]hen he perceives clearly and adequately that he is within nature, a part of its interactions, he sees that the line to be drawn is not between action and thought, or action and appreciation, but between blind, slavish, meaningless action and action that is free, significant, directed and responsible. (Dewey 1925/1981, 324)

Knowing what kinds of creatures we are, how our minds work, where our values come from, and how we appraise, evaluate, and remake experience are the keys to freeing up thought and making it intelligent. Our human animal minds are inextricably embodied and therefore tied to our visceral connection to the world. Whatever else your personal identity consists in, it cannot be separated from your brain, operating within your active, living body, as that body engages your material, interpersonal, cultural world. What and who we are has emerged in the ongoing evolutionary process (on the large scale), across centuries of historical cultural change (on the medium scale), and in our individual developmental processes (on the small scale) in which organisms engage their complex, changing environments and form themselves up in response to the contours, values, and possibilities for engagement afforded by those environments. That self is thus a self-in-process, until the moment you die. There is typically a good measure of stability in our identity over time, but it is only a pattern persisting through dynamic processes of change as we encounter our environment. Your moral selfhood resides not in your brain, not in your body, not in the world beyond your fleshy corpus, and not in your sociocultural reality, but rather in the ongoing interactions of all four of these dimensions of experience.

One important consequence of our embodied selfhood is that our human condition is finite, frail, fallible, and fearful, but it can also sometimes be exuberant, confident, joyful, celebratory, meaningful, and expansive. As we have noted repeatedly, we are creatures that need certain conditions to sustain our biological life, our interpersonal relations, our cultural institutions, and our sense of meaning and well-being. These conditions are indicative of the values we have. These values are natural (given that culture is a natural process), not supernatural or issuing from some alleged transcendent source. They emerge from our bodily, interpersonal, cultural engagement at multiple levels of organizational complexity.

Perhaps the single greatest obstacle to the acceptance of the view I am proposing is the necessity of abandoning the idea that "the right thing to do" exists antecedently in some Platonic moral space ready to be discovered through moral reasoning. There is no such preexistent moral essence, rule, or ideal to which we are supposed to match our concrete actions. When I concluded that moral fundamentalism is immoral, what I meant was that fundamentalism closes us off from the very forms of moral inquiry that are our only psychologically realistic hope for a better world. If you think you have the Absolute, One-and-Only Moral Truth, then why would you ever bother to seek an enriched understanding or to engage in problem-solving? Moral fundamentalism is a desperate and misguided attempt to overcome our human finitude and fallibility. It represents a quest to find an eternal, unchanging, unshakable foundation for living rightly and well. It is an attempt to escape from our human condition.

The alternative position that I have been developing is that morality is really a matter of moral inquiry, which is a form of problem-solving. Yes, there are intuitive, mostly nonconscious appraisals (Dewey's valuings) doing a great deal of the work of our moral assessment. There are also after-the-fact acts of reflective justification of our valuings, by which we shore up our intuitive judgments and try to defend our actions to ourself and others. To these two processes I have added moral deliberation as an ongoing activity of problem-solving that attempts to harmonize and enhance values from multiple domains of experience and multiple perspectives. This process of reflective valuation incorporates our intuitive processes of evaluation and assessment of situations, but those are primarily starting points for a process of imaginative dramatic rehearsal of possible courses of action open to us. The outcomes of such deliberative inquiries can be more and less reasonable, depending on how well they resolve the blockages and tensions that gave rise to our sense of moral conflict in the first place.

Conscientiousness is a primary virtue of moral inquiry, because it seeks the most comprehensive survey of conflicting ends, values, and principles, and it then strives for the most comprehensive resolution available within the situation as we currently understand it. It follows from this that we should appropriate all the relevant knowledge and understanding we are able to bring to bear to make sense of our problematic situation. It is here that we have to get beyond what we traditionally regard as "moral" knowledge and instead use every method of gaining understanding that is available to us. This will include all of the sciences, both "natural" (e.g., physics, chemistry, biology, neuroscience) and "social" (psychology, sociology, anthropology, linguistics, history), as well as various philosophical perspectives (phenomenology, existentialism, hermeneutics, pragmatism), and embracing also all of the arts and humanities as providing insight into the non-human and human worlds. It is in this sense, then, that morals ought to be seen as empirical, experimental, and experiential.

Success at moral problem-solving is a humbling activity that requires sensitivity, perceptiveness, comprehensiveness of thought, empathy, goodwill, a cooperative spirit, imagination, and sometimes a good measure of dumb luck. A sense of humor helps, too. Moral problem-solving is a transformation of the world, because it remakes experience. It is, at the same time, a self-transformation—a reconstitution of our identity, insofar as it reconfigures our habits and our relation to our surroundings, both physical and social. For each of us, it is never done until we die, simply because change is intrinsic to experience. So long as we live, life will present us with new challenges that cannot be adequately met with our currently sedimented habits, dispositions, values, and practices. So we need to cultivate the virtues of moral inquiry, develop flexibility of thought, and exercise our moral imagination.

This portrait of ongoing imaginative exploration and reconstruction need not induce in us a sense of failure or gloom stemming from an acknowledgment that we can never live up to what God, practical reason, society, or moral law demands of us. Morality is not about living up to such transcendent ideals. Rather, it is about humanly realistic ideals that grow out of our communal experience. Dewey captures the essence of this hope:

Because intelligence is critical method applied to goods of belief, appreciation and conduct, so as to construct freer and more secure goods, turning assent and assertion into free communication of shareable meanings, turning feeling into ordered and liberal sense, turning reaction into response, it is the reasonable object of our deepest faith and loyalty, the stay and support of all reasonable hopes. (1925/1981, 325)

An important part of moral growth is coming to see that morality is about *better and worse*—not about the Right or the Good—and cultivating our dispositions and skills for determining better and worse in actual human situations. For the fact is that sometimes things do actually develop for the better. Sometimes we manage to be perceptive and sensitive enough to achieve a breadth of perspective that encompasses much of what matters in a particular situation, and to see a way to harmonize competing interests and perspectives so that people feel more free, more connected, more respected, more fulfilled, more cared for, better understood, and more able to act constructively than before. And when that happens as a result of our moral deliberation, it is a consummation much to be desired, and it is often accompanied with a sense that grace was somehow at work—not the grace of an Almighty Father, but the harmonious convergence of various constituents of a world-in-process.

Acknowledgments

Five or six years ago, the Undergraduate Philosophy Club at the University of Oregon invited me to debate with my colleague Cheyney Ryan on the merits of naturalized approaches to ethics. That stimulating evening event was the inspiration for this book. In the following year, I received a fellowship from the Oregon Humanities Center to begin research on this project, which gave me an opportunity to immerse myself in the vast literature in this growing area of interest. Since that time, I have benefited immensely from extended conversations with Scott Pratt, especially during long drives up to the High Cascades fishing lakes, about our mutual interest in pragmatist conceptions of value. I am also indebted to Robert McCauley, who helped me craft the basic structure of my project in several conversations during a lovely week on the Oregon coast. A number of generous people read various drafts of this book and gave me most helpful advice. In particular, I would like to thank Owen Flanagan, John Kaag, Colin Koopman, Don Morse, Gregory Pappas, Jay Schulkin, and Richard Shusterman. Each of these folks was kind enough to see beyond the considerable shortcomings of the drafts they read and to help me say more clearly, fairly, and concisely what I was trying to explore about the natural origins of human value and deliberation. I am also grateful to Jeremy Swartz for his invaluable technical assistance in the preparation of this book, and to Caroline Lundquist for her preparation of the index.

Notes

INTRODUCTION

1. I am embracing Owen Flanagan's "Principal of Minimal Psychological Realism," which states, "Make sure when constructing a moral theory or projecting a moral ideal that the character, decision processing, and behavior are possible, or are perceived to be possible, for creatures like us" (Flanagan 1991, 32). I see little or no point in a moral philosophy that articulates values, principles, or virtues that are incompatible with what we have strong (if not overwhelming) evidence to believe are fundamental constraints on our cognition, emotional response, appraisal, motivation, decision processes, will, and so on.
2. I say this in jest, of course, since anyone traveling around the entire middle half of the United States—from Ohio in the east to Colorado in the west, from Minnesota in the north to Texas in the south—will hear many different places proudly proclaim that they are "the Heartland"—the true moral, spiritual, and political center of what it means to be an American.
3. Some folks seemed to prefer the popular baseball analogy, with its reference to "getting to first base," "getting to second base," and "getting to third base." But whether it was zones or bases, for many of us it was never completely clear whether third base was a term of achievement only in the petting sphere, or whether it went beyond that. And then there was the puzzling incompleteness of the base-running analogy, because nobody, as far as I can remember, ever referred to "getting to home plate."
4. The importance of such metaphorical motions—from *up* at the site of intelligence (the head) all the way *down* to the site

of procreation (the genitals), from *light* to *heavy* petting—was not missed by me, even back then before I had done thirty years of research on how metaphors structure our lives and values. There was a definite value-laden *up-down* orientation, with *down* carrying a decidedly negative connotation.

5. Kant's *Metaphysics of Morals* claims to offer a systematic treatment of the general duties that apply to rational creatures as such, at a level that abstracts away from anything particular to one person or another (such as age, gender, authority, social status). A metaphysics of morals "has to investigate the idea and principles of a possible pure will and not the actions and conditions of human volition as such, which are for the most part drawn from psychology" (Kant 1797/1983, 4).

6. I have long considered this part of Kant's moral philosophy a perfect example of a philosopher's believing they know in advance what the "right" moral view must be, and then finding a justification, however strained or implausible, to support their intuitions. This strikes me as a general human temptation and not just a preoccupation of philosophers.

7. Some might think it obvious what "sex outside marriage" means, but as we will see in a later chapter, the meaning of this phrase is anything but clear in the Hebrew and Christian scriptures. For example, does "sex" only refer to penetration of the vagina by a penis? Then, as has recently been reported in the news, "good" Christian girls can give their boyfriends blow jobs, as a way of satisfying them, without technically breaking the alleged moral prohibition of "sex" before marriage. Furthermore, consider what this penis-in-vagina definition of "sex" entails for adultery. Can I make out with my neighbor's wife, and that's okay, as long as I don't "lie" with her? I doubt my neighbor will think so, but what's to prevent this from being morally permissible, if sex means sexual intercourse?

8. Research on this pervasive form of after-the-fact moral justification is nicely summarized in Haidt (2012). There is growing evidence that people tend to make up even wildly implausible accounts of moral principles or values for moral intuitions for which they have no reasonable justifications.

9. This argument was the source of the popular antiwar bumper sticker that read "Fighting for Peace Is like Fucking for Chastity."

10. This interpretation of Rawls's project is evident in his three important essays organized under the general title "Kantian Constructivism in Moral Theory," *Journal of Philosophy* 77, no. 9 (1980): 515–72.

11. Emmanuel Levinas's emphasis on the "face of the other" is a classic example of this form of "argument" in non-naturalistic moral theory (Levinas 1969). The face of the other is supposed to confront us with an absolute limit to our actions, and to impose a set of unconditional moral obligations on us toward the other, namely, negatively, not to kill them (or do them harm), and positively, to feed them and aid them in the continuance of life. Our moral responsibility arising from our encounter with the other is a primordial moral reality, not subject to analysis or justification, because it

is not a theoretical claim at all, but rather an allegedly inescapable realization of an absolute moral relation.
12. The first two sections of Kant's *Grounding for the Metaphysics of Morals* (1785/1983) repeat *ad nauseum* that "the ground of obligation here must therefore be sought not in the nature of man nor in the circumstances of the world in which man is placed, but must be sought a priori solely in the concepts of pure reason . . ." (389).
13. For a spirited, deeply knowledgeable, and expansive view of Kant's system as acknowledging a place for more than pure practical reason, see Robert Louden's *Kant's Impure Ethics: From Rational Beings to Human Beings* (2000). Louden makes the strongest possible case for a more nuanced and empirically adequate interpretation of the Kantian orientation, but he does not deny Kant's insistence on the grounding of morality in pure reason.
14. See, for instance, Russ Shafer-Landau (2003, chap. 3), for a favorable assessment of Moore.
15. The best treatment I know concerning questions of will, choice, and responsibility from a naturalized perspective is Owen Flanagan's chapter "Free Will" in his book *The Problem of the Soul* (2002).

CHAPTER ONE

1. MacIntyre thus distinguishes between ends "internal" or "intrinsic" to a practice (such as the growing of delicious, healthy tomatoes) and ends "external" or "extrinsic" to the practice (such as the making of money by growing and then selling those tomatoes). Internal ends and virtues define the distinctive character of the practice, whereas externalities are simply add-ons that can apply to many different practices (e.g., various practices can make money), and so they do not define the essence of any particular practice. MacIntyre observes how easy it can be to get caught up with extrinsic ends and to miss the importance of the very intrinsic ends and virtues that are what the practice is really about. Notice how sportscasters are prone to want us and themselves to believe that, when professional sporting events come down to the playoffs and the finals, athletes are not "doing it for the money" but, rather, "for the love of the game" and the value of the excellences intrinsic to that particular type of sporting event.
2. Thus, practices like football, baseball, basketball, farming, and medicine undergo transformation over time, both in their constitutive rules and even sometimes in their conception of the ends and excellences internal to the practice.
3. I do not think it is reductivist to see moral knowledge as a form of technology, as long as that term is taken in its very broadest sense. If we can get beyond our overly narrow traditional model of technology as the application of bodily skills to shape materials for the realization of certain pre-given ends or states of affairs (as manipulation of means to realize ends),

we can understand technology in general as any intelligent, skillful means for transforming experience. This is the expansive sense in which Dewey understood the term: "'Technology' signifies all the intelligent techniques by which the energies of nature and man are directed and used in satisfaction of human needs; it cannot be limited to a few outer and comparatively mechanical forms" (Dewey 1930b/1988, 270). Isn't this precisely what good moral reasoning does; namely, it brings human intelligence skillfully to bear on a process by which experience is changed for the better, thereby at least temporarily "solving" a particular problem of how to behave. Many people recoil at the suggestion that moral reasoning is problem-solving, primarily because they think of problem-solving as solely a means-to-the-achievement-of-given-ends process. However, the fact that some types of problem-solving are crude and simplistic calculations of means to pregiven ends does not mean that all problem-solving reduces to this kind of reasoning. Dewey did not mean by "satisfaction" only that our ends are always given and then technology helps us achieve (i.e., satisfy) those ends. Rather, there are also technologies in which the ends are not determinately specified in advance, and where the ends-in-view can change through the operation of an intellectual technology to help us work through a problematic situation, transforming it for the better. Larry Hickman (1990) has championed this broad, expansive reading of Dewey's treatment of technology as incorporating methods of intelligent inquiry and problem-solving in all domains of human life.

4. I do not agree with Hauser's account of the father's response as being "the same" in both cases, but I will accept his description for purposes of setting out his underlying argument.
5. This point is made quite convincingly by physician and philosopher Gary Wright in his book *Means, Ends and Medical Care* (2007).
6. Kant writes, "All judgments are functions of unity among our representations. . . . Now we can reduce all acts of understanding to judgments . . ." (Kant 1781/1968, A69/B93–94).
7. Kant's "argument" for this claim about pure moral philosophy rests on his further assuming a set of distinctions about types of judgments (empirical vs. pure) and modes of knowledge (*a priori* vs. *a posteriori*) that have come under withering attack from quite different philosophical orientations over most of the twentieth century.
8. This interpretation of Kant as giving a rational reconstruction of his particular moral tradition can be found in Rorty (1979), MacIntyre (1984), Johnson (1993), Putnam (2004), and many others.
9. For the role of empirical considerations in Kant's moral theory, see Gregor (1963). Robert Louden (2000) also does an outstanding job of describing the place of empirical knowledge in what he calls Kant's "impure philosophy." In his *Metaphysics of Morals*, Kant describes the process of bringing increasingly empirical considerations into play in order to determine what

moral laws prescribe for our behavior. He mentions an "appendix" to a pure metaphysics of morals, which involves "applying pure principles of duty to cases of experience so as to schematize them, as it were, and to set them forth ready for morally-practical use" (1797/1983, 468). However, every addition of empirical knowledge, according to Kant, would move us ever further from the pure moral philosophy that is "indispensably necessary."

10. Many would argue that we should not limit our respect only to what Kant called "rational creatures," which, for Kant, excluded other animals, and perhaps even some human beings. Some would urge us to extend the notion of dignity, as a basis for moral regard, to many or all animals, and even to non-sentient systems, such as ecosystems.

11. Alan Donagan (1977), Robert Louden (2000), and many others have recognized a role for empirical knowledge in Kant's moral theory, but that knowledge is relevant primarily in the application of moral principles to concrete situations, and *not* in the formation of the fundamental moral principles themselves.

12. "Moreover, one cannot better serve the wishes of those who ridicule all morality as being a mere phantom of human imagination ... than by conceding to them that the concepts of duty must be drawn solely from experience," and "worse service cannot be rendered morality than that an attempt be made to derive it from examples" (Kant 1785/1983, 407, 408).

CHAPTER TWO

1. I put scare quotes around "normal" to emphasize that the normal is nothing but a statistical frequency of developmental patterns within a species and *not* a rigid and uniform universal sequence of developmental stages.

2. Again, let me emphasize that "typical" here means only that a very large percentage of a species or group would traverse the same developmental stages.

3. These levels do not represent discrete, independent dimensions of moral experience. They are quite obviously intertwined, such as when virtues that arise in the context of interpersonal intimate relations (e.g., mother-child transactions) are equally relevant to higher-level social organization (e.g., membership and participation in a community health project).

4. Of course, there are some interpersonal and group relationships that we do choose (e.g., Girl Scouts; whether to affiliate with any church, synagogue, or mosque; social clubs; sports teams, labor unions, etc.). My point is that (1) we are by nature social persons, and (2) there are obviously certain personal relations and social groups that we simply find ourselves a part of and do not, in any sense, *choose*.

5. I make this universal care claim with the clear understanding that the relevant "others" that are to be cared for can be quite differently defined

NOTES TO PAGES 60–75

      across cultures. Furthermore, what constitutes a minimal level of care can also vary across different societies and subgroups.

6. Sarah LaChance Adams (2011) challenges various overly romanticized conceptions of mothering, but she does this to argue for a more psychologically realistic conception of an ethics of care.
7. The currently most popular hypothesis on offer is that it is precisely an inability to experience empathy that characterizes the sociopathic personality, for whom the other can become a mere thing to be used for the satisfaction of one's desires and purposes.
8. It is this woeful absence of civic-mindedness among members of the U.S. Congress that is today threatening to destroy our democratic system of government.
9. This praise of rationality and intelligence is put forward while at the same time acknowledging the vast mess we humans seem to have made of our social and political relations, our environment, and our intimate relations. I claim only that we are better off, by virtue of our capacity for reasoning and problem-solving, than if we lacked those capacities. Although I share Dewey's faith in human intelligence as our best hope for a better future, I also share his recognition that reason alone (if there is such a thing) is inadequate to the task without aesthetic and moral sensibility, proper education, minimal material resources, good will toward others, and a host of other virtues and basic conditions for flourishing.
10. One key disagreement, for example, concerns Haidt's (2012) hypothesis that contemporary conservative political systems incorporate all five of the foundations, while liberal or progressive systems tend to recognize only two (care/harm and fairness/cheating). Haidt speculates that this alleged impoverishment of liberal moralities explains the intuitive appeal to so many of conservation perspectives, which supposedly embrace a broader spectrum of human concerns. I would simply observe that Lakoff's analysis of what he calls "Strict Parent" moral systems versus "Nurturant Parent" moralities and political orientations reveals components of all six of Haidt's foundations as present in Nurturant Parent–based progressive views (Lakoff 2002, 2004, 2008).

CHAPTER THREE

1. Most religious perspectives also grant moral status to angels or other disembodied spirits, but in such cases it is typically their possession of autonomous reason and will that justifies their being regarded as ethical beings. On the other end of the species scale, when non-human animals are granted moral status, it is usually on the basis of arguments that they possess cognitive capacities sufficiently similar to those of humans to merit the attribution of a free reason and will.
2. Hauser's description of the twin processes closely parallels Haidt's: "Intui-

tions are fast, automatic, involuntary, require little attention, appear early in development, are delivered in the absence of principled reasons, and often appear immune to counter-reasoning. Principled reasoning is slow, deliberate, thoughtful, requires considerable attention, appears late in development, justifiable, and open to carefully defended and principled counterclaims" (Hauser 2006, 31).
3. Neither, as I will argue later, are we strictly Rawlsian creatures, as Hauser claims, because Rawls did not fully appreciate the crucial role of emotions in moral judgment.
4. Haidt and Kesebir (2010) have nicely summarized ten key types of research evidence that supports the importance of intuitive processes in our thinking: (1) People make rapid evaluative judgments of others; (2) moral judgments involve brain areas related to emotion; (3) morally charged economic behaviors involve brain areas related to emotions; (4) psychopaths have emotional deficits; (5) moral-perceptual abilities emerge in infancy; (6) manipulating emotions changes judgments; (7) people sometimes can't explain their moral judgments; (8) reasoning is often guided by desires; (9) research in political psychology points to intuitions, not reasoning; (10) research on prosocial behavior points to intuitions, not reasoning. Haidt and Kesebir coin the phrase "Intuitive Primacy (but not Dictatorship)" to capture their idea that although most of our moral thinking is an intuitive process, there remains a place for more conscious, deliberative, reflective thinking processes of evaluation and judgment too.
5. Recall, from chapter 2, that I am typically replacing "homeostasis" (which involves a return to some set-point equilibrium) with "allostasis" (as defined in Schulkin [2011]), which captures the more dynamic character of the equilibrium, in which new states of balance can emerge, subject to subsequent recalibration.
6. "My Hypothesis, then, presented in the form of a provisional definition is that *a feeling is the perception of a certain state of the body along with the perception of a certain mode of thinking and of thoughts with certain themes*," and "feelings are functionally distinctive because their essence consists of the thoughts that represent the body involved in a reactive process" (Damasio 2003, 86; italics in original).
7. Damasio (1994, 1999, 2003) develops an elaborate account of why and how humans have feelings. Fortunately, for my purposes, I do not have to set out his theory of feeling, since I need only the acknowledgment that we do feel the types of processes we are describing—especially the tension, blockage, flow, and surge of our life processes.
8. The central role of affect in moral cognition leads Damasio to conclude that without a capacity to have certain emotions and to feel certain emotions, morality would never have come into existence. He notes that there is often profound social and moral dysfunction in patients who suffer certain types of prefrontal cortex damage very early in life, which makes it

impossible for them to connect emotional responses with morally charged situations. He then asks us to imagine a world in which everybody was this way naturally, and therefore nobody had acquired the ability to respond to other people with sympathy, gratitude, shame, guilt, contempt, embarrassment, or any of the social emotions. He correctly observes that people in such a world would have nothing resembling human morality, because human morality is grounded in our emotional responses to our world and to other people. He describes the amoral world that would result:

> In a society deprived of such emotions and feelings, there would have been no spontaneous exhibition of the innate social responses that foreshadow a simple ethical system—no budding altruism, no kindness when kindness is due, no censure when censure is appropriate, no automatic sense of one's own failings. In the absence of the feelings of such emotions, humans would not have engaged in a negotiation aimed at finding solutions for problems faced by the group, e.g., identification and sharing of food resources, defense against threats or disputes among its members. There would not have been a gradual building up of wisdom regarding the relationships among social situations, natural responses, and a host of contingencies such as the punishment or reward incurred by permitting or inhibiting natural response. The codification of rules eventually expressed in systems of justice and sociopolitical organizations is hardly conceivable in those circumstances . . . (2003, 157).

CHAPTER FOUR

1. Dewey is not alone in emphasizing the profound importance of the larger experiential context in which all our thought and action arises. Although they have different interpretations and focal interests with regard to the environing situation, many philosophers have recognized that only within an experiential situation do any objects, events, and meanings stand forth for us. Husserl speaks of the "lifeworld," Heidegger of the "world," Merleau-Ponty of the "flesh of the world," Searle of the "Background," and Wittgenstein gestures toward the primacy of the situation within his notion of language games. Among these and others, Husserl, Heidegger, and Dewey seem to have appreciated best the full significance of the qualitative in the constitution and development of a situation.
2. "*An* object or event is always a special part, phase, or aspect, of an environing experienced world—a situation. The singular object stands out conspicuously because of its especially focal and crucial position at a given time in determination of some problems of use or enjoyment which the *total* complex environment presents" (Dewey 1938/1991, 72; italics in original).
3. This practice of reading our theoretical abstractions and methodological assumptions back into what we take to be moral experience is what William

James called "the Psychologist's Fallacy": "The *great* snare of the psychologist is the *confusion of his own standpoint with that of the mental fact* about which he is making his report" (1890/1950, vol. I, 196; italics in original).

4. Artworks are not just *things*. They are events as experienced. They exist only as enactments of meaning for an observer in a sociocultural context. Therefore, the experience of an artwork is an enactment of a particular *situation*, and that is why Dewey can use artworks to illustrate his notions of a situation and the pervasive unifying quality of a situation. See Dewey (1934/1987).

5. See John Kaag and Sarah Kreps (2012) for an in-depth exploration of these issues.

6. Virtually all of Dewey's major works have detailed and insightful accounts of the roots of significant moral, political, and intellectual problems. I am thinking here of works such as *Human Nature and Conduct* (1922), *Experience and Nature* (1925), *The Public and Its Problems* (1927), *The Quest for Certainty* (1929), *Individualism Old and New* (1930), and *Logic: The Theory of Inquiry* (1938).

7. See also Feldman (2006) and Bergen (2012) for a broad outline of a simulation semantics.

8. Concepts are not defined by sensory and motor simulations alone, obviously, since there will also be cultural dimensions involved as part of our simulations. For example, part of our concept of a car will include typical social situations surrounding particular brands and models, such as social status associated with certain types of vehicles, and the kinds of social situations in which cars figure prominently. There will also be possible economic, political, religious, and aesthetic considerations in play in running a conceptual simulation.

CHAPTER FIVE

1. Dewey's *Logic: The Theory of Inquiry* (1938) is a massively sustained argument for the primacy of activities of inquiry as the basis for logic, rather than articulating logic prior to any particular pattern of reasoning. Logical principles are constraints that have proved useful to communities of inquirers for conducting certain specific types of inquiries, so logic is not a pre-given absolute structure of right thinking. It emerges as patterns of successful investigation and inquiry. It is arrived at through experience, not discovered as if one had uncovered the mother lode of rationality.

2. Annette Baier (1991) makes a strong case for Hume as proposing a more nuanced view of reason as blended with sentiment and shaped by social practices—a view quite close to the one I am developing here. See also Cohon (2008), who questions the traditional interpretation of Hume as placing all the weight on moral emotion, to the exclusion of reason.

CHAPTER SIX

1. This current chapter is taken from Johnson (2012), with some additions and deletions. A more multidimensional and modest version of the moral instinct position is Jonathan Haidt and Craig Joseph's "modified nativist view" that there exists an "*intuitive ethics*, an innate preparedness to feel flashes of approval or disapproval toward certain patterns of events involving other human beings" (2004, 56). What in Hauser was a single faculty is replaced in Haidt and Joseph (and in Haidt 2012) by a more nuanced set of four or five moral modules—deep cognitive/affective structures that generate predictable patterns of judgment and response concerning things like suffering, hierarchy relations, reciprocity among persons, and notions of moral purity. What is attractive about this view is the way it identifies a small number of factors or dimensions that tend to be found in moral systems around the world and throughout history. However, for reasons articulated below, one must remain cautious about *any* notion of a moral module.
2. The article on which this chapter is based was written prior to the emergence of questions about some of Hauser's academic practices. I take no position on that controversy, and I would not want my criticism of Hauser to be viewed as in any way piggybacking on the general academic criticisms currently circulating. On the contrary, I find Hauser's book (2006) to be a major statement of the moral grammar perspective that merits our attention for its significant contribution to the debate over how to account for what appear to be universally shared aspects of moral systems. Moreover, the parts of Hauser's work with which I take issue do not rest on any of the experimental work that has been called into question by some critics.
3. Mallon bases his criticism of Hauser's moral faculty view on the argument that there may be multiple neural realizations of any set of alleged principles of the sort envisioned by Hauser: "Most simply, we can imagine that a correct and complete description of our moral capacity (Level 1) invokes simple moral principles that are literally represented in the brain and used in computations to generate moral judgment (Level 2), and this is all carried out in a particular way by a particular, functionally distinct brain region (Level 3). However, an equally familiar point from these discussions in cognitive science is that properties functionally specified at a higher level of description may be realized by a variety of different lower level mechanisms. Thus, the mere fact that we can describe specific jobs for a moral faculty (e.g., action appraisal) ought to give us no confidence at all that there really is some specialized computational faculty (Level 2) or brain region (Level 3) that realizes such a function" (2008, 150).
4. I hold the opinion that a deep motivation for the return to moral faculty theories is a profound desire to counter moral relativism, at least in its extreme forms. Whatever else moral instinct/faculty theories claim, they all

insist that there are at least some universals underlying cultural variations. On this view, the buck has to stop somewhere, so that at least something is fixed and final, even if it is a very abstract principle. We have a nagging desire for foundations. Hauser and others hold the "principles and parameters" view, according to which there are universal principles, with, however, those principles containing parameters that can be differently specified by different cultures. In this way, we supposedly can retain a small measure of universality, while recognizing a decent measure of cultural variability in moral practices, values, and principles. Thus Dwyer, Huebner, and Hauser assert that the linguistic analogy "predicts that there are invariant, universal principles that lie at the core of the FM [moral faculty], perhaps with parameters providing options to create cross-cultural variation" (2010, 501).

5. To repeat my earlier point (chapter 1), I am not denying the commonplace that people tend to call some situations and issues "moral," some "political," some "aesthetic," and some "scientific," to name just a few common discriminations we make. We call a situation "moral" when it seems prominently to involve matters of human well-being and responsible action. We call a situation "aesthetic" when we are interested primarily in its felt qualities and meaning. However, experiences don't come labeled in discrete types, and every so-called "moral situation" is often permeated with aesthetic, practical, technical, political, economic, and religious dimensions.

6. My opinion is that this is precisely what Hauser has done, that is, give the name of a faculty to a large group of interacting functional systems. On my reading, Hauser's most significant accomplishment lies in identifying (and describing) many of the systems that have to be intact for moral cognition, rather than in circumscribing an alleged faculty.

CHAPTER SEVEN

1. Robert McCauley (2011) argues that part of the nearly universal appeal of religions is that they characteristically use forms of explanation that draw on notions of agency, intentionality, and purposiveness. These forms of explanation intuitively seem to make sense to most people, because human agency is our first-learned mode for explaining what happens. A world in which things happen as a result of the intentional, intelligent agency of some rational being becomes a prototype for us of comprehensible causation.

2. A good representative example of how finite human beings can have concepts of infinity, without themselves needing to have some connection to something infinite, is found in Lakoff and Núñez (2000), where they analyze the different metaphors underlying our conceptions of actual and potential infinity in mathematics.

CHAPTER EIGHT

1. I am thinking of Richard Rorty's contrast between truth as *found* and truth as *made*—a distinction he readily transfers to our moral concepts. See Rorty (1989, 3).
2. This is what Dewey meant when he said, "But in fact morals is the most humane of all subjects. It is that which is closest to human nature; it is ineradicably empirical, not theological nor metaphysical nor mathematical. Since it directly concerns human nature, everything that can be known of the human mind and body in physiology, medicine, anthropology, and psychology is pertinent to moral inquiry. . . . Hence physics, chemistry, history statistics, engineering science, are a part of disciplined moral knowledge so far as they enable us to understand the conditions and agencies through which man lives, and on account of which he forms and executes his plans. Moral science is not something with a separate province. It is physical, biological, and historic knowledge placed in a human context where it will illuminate and guide the activities of man" (1922, 204–5).
3. Perhaps Nietzsche is the most notable exception here, since he focuses primarily on growth of power or force in the organism as the basis for good. However, Nietzsche seems not to have Dewey's highly nuanced sense of what growth consists in, preferring as he does the man of strength (*virtue* and manliness) who spontaneously wills his own good out of the power that is within him.
4. I have attempted to articulate Dewey's pragmatist theory of meaning in Johnson (2007), as follows: "Human meaning concerns the character and significance of a person's interactions with their environments. The meaning of a specific aspect or dimension of some ongoing experience is that aspect's connections to other parts of past, present, or future (possible) experiences. Meaning is relational. It is about how one thing relates to or connects with other things. The pragmatist view of meaning says that the meaning of a thing is its consequences for experience—how it 'cashes out' by way of experience, either actual or possible experience" (10).
5. The list of passages in the Hebrew scriptures indicating polygyny is far too extensive to bother listing. One need simply Google "polygyny in the Bible" to find passage after passage of this type of marital arrangement.
6. In the *Institutes*, for example, Calvin denies that marriage is a sacrament, and this remains the dominant Protestant understanding today.
7. I mention the key role of "good luck," because for many couples, getting pregnant is one of the hardest tasks they will ever face.
8. It is important to note that denying the use of the term "natural" as a basis for moral evaluation does not preclude reference to norms, either in the descriptive sense (as what a group tends to do or to favor, from a statistical viewpoint) or in the evaluative sense (as what a group approves of as de-

cent behavior and social standards). There is thus no need for a metaphysically natural grounding in order to engender and support group norms.
9. See, for example, Maureen Sullivan, *The Family of Woman: Lesbian Mothers, Their Children, and the Undoing of Gender* (2004).
10. In challenging the moral legitimacy of certain historically established purification rights, I do not thereby deny the importance of purity conditions (physical, moral, and spiritual) in religions and cultures around the world and throughout history. I acknowledge Haidt's (2012) analysis of purity as a fundamental dimension of most moral systems, but this need not prevent us from rejecting certain practices in light of our current scientific understanding of human development, social interaction, and cultural practice.
11. Kierkegaard was right when he argued that literal adherence to revealed truths and commandments of a holy God is not for the weak of heart, and he was also right to observe that the Hebrew and Christian faith cannot be made comfortably rational (as witnessed, for example, by the infamous story of Abraham's willingness to sacrifice his son Isaac at God's commandment). But Kierkegaard is also right when he observes, with horror, that there can be no certain way to determine whether the voice of God commanding Abraham's terrible sacrificial deed is not perhaps the voice of Satan or the delusional result of some fundamental psychopathology. I confess that Abraham's willingness to cut his beloved son's throat and burn his body (with wood he has required that same son to carry on his back to the mountain!) as an offering to an all-demanding God evokes horror in me.
12. There are numerous outstanding accounts of the ways in which literary fictions can be morally enlightening because narrative provides us the critical context and details of the moral situations we encounter, and it meshes with the narrative structure of our self-understanding (Eldridge 1989; Ricoeur 1992; Johnson 1993; Gregory 2009).

References

Aziz-Zadeh, Lisa, S.M. Wilson, Giacomo Rizzolatti, and Marco Iacobini. 2006. "Congruent Embodied Representations for Visually Presented Actions and Linguistic Phrases Describing Actions." *Current Biology* 16 (18): 1818–23.

Baier, Annette. 1991. *A Progress of the Sentiments: Reflections on Hume's "Treatise."* Cambridge: Cambridge University Press.

Barsalou, Lawrence. 1999. "Perceptual Symbol Systems." *Behavioral and Brain Science* 22: 577–660.

———. 2003. "Situated Simulation in the Human Conceptual System." *Language and Cognitive Processes* 18 (5/6): 513–562.

Bechtel, William. 2009. *Mental Mechanisms: Philosophical Perspectives on Cognitive Neuroscience.* New York: Psychology Press.

Bergen, Benjamin. 2012. *Louder than Words: The New Science of How the Mind Makes Meaning.* New York: Basic Books.

Casebeer, William. 2003. *Natural Ethical Facts: Evolution, Connectionism, and Moral Cognition.* Cambridge, MA: MIT Press.

Churchland, Patricia S. 2011. *Braintrust: What Neuroscience Tells Us about Morality.* Princeton, NJ: Princeton University Press.

Churchland, Paul M. 2007. "Toward a Cognitive Neurobiology of the Moral Virtues." In *Neurophilosophy at Work,* 37–60. Cambridge: Cambridge University Press.

Cohon, Rachel. 2008. *Hume's Morality.* Oxford: Oxford University Press.

Damasio, Antonio. 1994. *Descartes' Error: Emotion, Reason, and the Human Brain.* New York: G. P. Putnam's Sons.

———. 1999. *The Feeling of What Happens: Body and Emotion in the Making of Consciousness.* New York: Harcourt Brace.

———. 2003. *Looking for Spinoza: Joy, Sorrow, and the Feeling Brain.* Orlando, FL: Harcourt.

———. 2010. *Self Comes to Mind: Constructing the Conscious Brain.* New York: Pantheon.

de Waal, Frans. 1996. *Good Natured: The Origins of Right and Wrong in Humans and Other Animals*. Cambridge, MA: Harvard University Press.

———. 2008. "Putting the Altruism Back in Altruism: The Evolution of Empathy." *Annual Review of Psychology* 59: 279–300.

Decety, Jean, and Julie Grezes. 2006. "The Power of Simulation: Imagining One's Own and Other's Behavior." *Brain Research* 1079: 4–14.

Decety, Jean, and Philip Jackson. 2004. "The Functional Architecture of Human Empathy." *Behavioral and Cognitive Neuroscience Reviews* 3: 71–100.

Dewey, John. 1922. *Human Nature and Conduct*. Vol. 14 of *The Middle Works, 1899–1924*. Edited by Jo Ann Boydston. Carbondale: Southern Illinois University Press.

———. 1925/1981. *Experience and Nature*. Vol. 1 of *The Later Works, 1925–1953*. Edited by Jo Ann Boydston. Carbondale: Southern Illinois University Press.

———. 1930a/1988. "Qualitative Thought." In Vol. 5 of *The Later Works, 1925–1953*. Edited by Jo Ann Boydston. Carbondale: Southern Illinois University Press.

———. 1930b/1988. "What I Believe." In Vol. 5 of *The Later Works, 1925–1953*. Edited by Jo Ann Boydston. Carbondale: Southern Illinois University Press.

———. 1934/1987. *Art as Experience*. Vol. 10 of *The Later Works, 1925–1953*. Edited by Jo Ann Boydston. Carbondale: Southern Illinois University Press.

———. 1938/1991. *Logic: The Theory of Inquiry*. Vol. 12 of *The Later Works, 1925–1953*. Edited by Jo Ann Boydston. Carbondale: Southern Illinois University Press.

———. 1939/1991. *Theory of Valuation*. Vol. 13 of *The Later Works, 1925–1953*. Edited by Jo Ann Boydston. Carbondale: Southern Illinois University Press.

Dewey, John, and James H. Tufts. 1932/1989. *Ethics*. Vol. 7 of *The Later Works, 1925–1953*. Edited by Jo Ann Boydston. Carbondale: Southern Illinois University Press.

Donagan, Alan. 1977. *The Theory of Morality*. Chicago: University of Chicago Press.

Dwyer, Susan, Bryce Huebner, and Marc Hauser. 2010. "The Linguistic Analogy: Motivations, Results, and Speculations." *Topics in Cognitive Science* 2: 486–510.

Eagleton, David. 2001. *Incognito: The Secret Lives of the Brain*. New York: Random.

Edel, Abraham. 2001. "Nature and Ethics." In *Encyclopedia of Ethics*, 2nd ed. Edited by L. Becker and C. Becker, 1217–20. London: Routledge.

Edelman, Gerald. 1992. *Bright Air, Brilliant Fire: On the Matter of Mind*. New York: Basic Books.

———. 2004. *Wider than the Sky: the Phenomenal Gift of Consciousness*. New Haven, CT: Yale University Press.

Edelman, Gerald, and Giulio Tononi. 2000. *A Universe of Consciousness: How Matter Becomes Imagination*. New York: Basic Books.

Eldridge, Richard. 1989. *On Moral Personhood: Philosophy, Literature, Criticism, and Self-Understanding*. Chicago: University of Chicago Press.

Feldman, Jerome. 2006. *From Molecule to Metaphor: A Neural Theory of Language.* Cambridge, MA: MIT Press.

Fesmire, Steven. 2003. *John Dewey and Moral Imagination: Pragmatism in Ethics.* Bloomington: Indiana University Press.

Feyerabend, Paul. 1975. *Against Method.* London: New Left Books.

Fillmore, Charles. 1985. "Frames and the Semantics of Understanding." *Quaderni di Semantica* 6: 222–53.

Flanagan, Owen. 1991. *Varieties of Moral Personality: Ethics and Psychological Realism.* Cambridge, MA: Harvard University Press.

———. 1996. "Ethics Naturalized: Ethics as Human Ecology." In *Mind and Morals: Essays on Ethics and Cognitive Science,* edited by Larry May, Marilyn Friedman, and Andy Clark, 19–44. Cambridge, MA: MIT Press.

———. 2002. *The Problem of the Soul: Two Visions of Mind and How to Reconcile Them.* New York: Basic Books.

———. 2007. *The Really Hard Problem: Meaning in a Material World.* Cambridge, MA: MIT Press.

Flanagan, Owen, and Robert Williams. 2010. "What Does the Modularity of Morals Have to Do with Ethics?: Four Moral Sprouts Plus or Minus a Few." *Topics in Cognitive Science* 2: 430–53.

Forceville, Charles, and Eduardo Urios-Aparisi. 2009. *Multimodal Metaphor.* Berlin: Mouton de Gruyter.

Frankl, Viktor. 1946/2006. *Man's Search for Meaning.* Boston: Beacon Press.

Freud, Sigmund. 1927/1968. *The Future of an Illusion.* In *The Standard Edition of the Complete Psychological Works of Sigmund Freud.* Edited by J. Strachey. London: Hogarth Press.

Gable, Shelly, and Jonathan Haidt. 2005. "What (and Why) Is Positive Psychology?" *Review of General Psychology* 9 (2): 103–10.

Gallagher, Shaun, and Daniel Hutto. 2008. "Understanding Others through Primary Interaction and Narrative Practice." In *The Shared Mind: Perspectives on Intersubjectivity,* edited by J. Zlatev, T. Razine, C. Sinha, and E. Itkonen, 17–38. Amsterdam: John Benjamins.

Gallese, Vittorio. 2001. "The Shared Manifold Hypothesis: From Mirror Neurons to Empathy." *Journal of Consciousness Studies* 8: 33–50.

Gallese, Vittorio, Luciano Fadiga, Leonardo Fogassi and Giacomo Rizzolatti. 1996. "Action Recognition in the Premotor Cortex." *Brain* 119(2): 593–609.

Geertz, Clifford. 1966. "Religion as a Cultural System." In *Anthropological Approaches to the Study of Religion,* edited by Michael Banton, 3–43. London: Tavistock.

Gendlin, Eugene. 1997. "How Philosophy Cannot Appeal to Experience, and How It Can." In *Language beyond Postmodernism: Saying and Thinking in Gendlin's Philosophy,* edited by M. Levin, 3–41. Evanston, IL: Northwestern University Press.

Gibbs, Raymond. 1994. *The Poetics of Mind: Figurative Thought, Language, and Understanding.* Cambridge: Cambridge University Press.

———. 2006. *Embodiment and Cognitive Science*. Cambridge: Cambridge University Press.

Gibson, James J. 1979. *The Ecological Approach to Visual Perception*. Boston: Houghton Mifflin.

Gilligan, Carol. 1982. *In a Different Voice: Psychological Theory and Women's Development*. Cambridge, MA: Harvard University Press.

Goldberg, Adele. 1995. *Constructions: A Construction Grammar Approach to Argument Structure*. Chicago: University of Chicago Press.

———. 2006. *Constructions at Work: The Natural of Generalizations in Language*. Oxford: Oxford University Press.

Goldman, Alvin. 1989. "Interpretation Psychologized." *Mind and Language* 4: 161–85.

Gopnik, Alison, and Andrew Meltzoff. 1997. *Words, Thoughts, and Theories*. Cambridge, MA: MIT Press.

Gordon, Robert. 1986. "Folk Psychology as Simulation." *Mind and Language* 1: 158–71.

Gregor, Mary. 1963. *Laws of Freedom: A Study of Kant's Method of Applying the Categorical Imperative in the "Metaphysik der Sitten."* Oxford: Basil Blackwell.

Gregory, Marshall. 2009. *Shaped by Stories: The Ethical Power of Narratives*. South Bend, IN: Notre Dame University Press.

Haidt, Jonathan. 2001. "The Emotional Dog and Its Rational Tail: A Social Intuitionist Approach to Moral Judgment." *Psychological Review* 108: 814–34.

———. 2006. *The Happiness Hypothesis: Finding Modern Truth in Ancient Wisdom*. New York: Basic Books.

———. 2007. "The New Synthesis in Moral Psychology." *Science* 316: 998–1002.

———. 2012. *The Righteous Mind: Why Good People Are Divided by Politics and Religion*. New York: Pantheon.

Haidt, Jonathan, and Craig Joseph. 2004. "Intuitive Ethics: How Innately Prepared Intuitions Generate Culturally Variable Virtues." *Daedalus* (Fall): 55–66.

Haidt, Jonathan, and Selin Kesebir. 2010. "Morality." In *Handbook of Social Psychology*, 5th ed., edited by S. Fiske, D. Gilbert, and G. Lindzey, 797–832. Hoboken, NJ: Wiley.

Hampton, Jean. 1993. "Feminist Contractarianism." In *A Mind of One's Own: Feminist Essays in Reason and Objectivity*, edited by L. Anthony and C. Witt, 227–55. Boulder, CO: Westview.

Harman, Gilbert. 2008. "Using a Linguistic Analogy to Study Morality." In *Moral Psychology*, Vol. 1: *The Evolution of Morality: Adaptations and Innateness*, edited by Walter Sinnott-Armstrong, 345–51. Cambridge, MA: MIT Press.

Hauser, Marc. 2006. *Moral Minds: How Nature Designed Our Universal Sense of Right and Wrong*. New York: HarperCollins.

Hauser, Marc, Liane Young, and Fiery Cushman. 2008. "Reviving Rawls's Linguistic Analogy: Operative Principles and the Causal Structure of Moral Actions." In *Moral Psychology*, Vol. 2: *The Cognitive Science of Morality: Intuition*

and Diversity, edited by Walter Sinnott-Armstrong, 107–43. Cambridge, MA: MIT Press.
Held, Virginia. 2005. *The Ethics of Care*. Oxford: Oxford University Press.
Henrich, Joe, Steve Heine, and Ara Norenzayan. 2010. "The Weirdest People in the World?" *Behavioral and Brain Sciences* 33: 61–83.
Hickman, Larry. 1990. *John Dewey's Pragmatic Technology*. Bloomington: Indiana University Press.
Hill, Thomas, Jr. 1980. "Humanity as an End in Itself." *Ethics* 91 (1): 84–99.
Hinde, Robert. 2002. *Why Good Is Good: The Sources of Morality*. London: Routledge.
Hoffman, Martin. 1987. "The Contribution of Empathy to Justice and Moral Judgment." In *Empathy and Its Development*, edited by N. Eisenberg and J. Strayer, 47–80. New York: Cambridge University Press.
———. 2000. *Empathy and Moral Development: Implications for Caring and Justice*. Cambridge: Cambridge University Press.
Hume, David. 1739/1888. *A Treatise of Human Nature*. Edited by L. A. Selby-Bigge. Oxford: Clarendon Press.
———. 1777. *An Enquiry concerning the Principles of Morals*. LaSalle, IL: Open Court, 1966.
James, William. 1890/1950. *The Principles of Psychology*. 2 vols. New York: Dover.
Johnson, Mark. 1987. *The Body in the Mind: The Bodily Basis of Meaning, Imagination, and Reason*. Chicago: University of Chicago Press.
Johnson, Mark. 1993. *Moral Imagination: Implications of Cognitive Science for Ethics*. Chicago: University of Chicago Press.
———. 2007. *The Meaning of the Body: Aesthetics of Human Understanding*. Chicago: University of Chicago Press.
———. 2012. "There Is No Moral Faculty." *Philosophical Psychology* 25 (3): 409–32.
Kaag, John, and Sarah Kreps. 2012. "The Use of Unmanned Aerial Vehicles in Asymmetric Conflict: Legal and Moral Implications." *Polity* 44 (2): 260–85.
Kant, Immanuel. 1781/1968. *Critique of Pure Reason*. Translated by N. K. Smith. New York: St. Martin's Press.
———. 1785/1983. *Grounding for the Metaphysics of Morals*. Translated by J. Ellington. In *Ethical Philosophy*. Edited by W. Wick. Indianapolis: Hackett.
———. 1787/1956. *Critique of Practical Reason*. Translated by L. W. Beck. Indianapolis: Bobbs-Merrill.
———. 1790/1987. *Critique of Judgment*. Translated by W. Pluhar. Indianapolis: Hackett.
———. 1797/1983. *Metaphysics of Morals*. Translated by J. Ellington. In *Ethical Philosophy*. Edited by W. Wick. Indianapolis: Hackett.
———. 1930/1963. *Lectures on Ethics*. Translated by L. Infield. Indianapolis: Hackett.
Kitcher, Philip. 2011. *The Ethical Project*. Cambridge, MA: Harvard University Press.

Kittay, Eva. 1999. *Love's Labor: Essays on Women, Equality and Dependency.* New York: Routledge.

Kohlberg. Lawrence. 1981. *Essays on Moral Development,* Vol. 1: *The Philosophy of Moral Development.* New York: Harper & Row.

———. 1984. *Essays on Moral Development,* Vol. 2: *The Psychology of Moral Development.* New York: Harper & Row.

Koopman, Colin. 2009. *Pragmatism as Transition: Historicity and Hope in James, Dewey, and Rorty.* New York: Columbia University Press.

Kovecses, Zoltan. 2010. *Metaphor: A Practical Introduction.* 2nd ed. Oxford: Oxford University Press.

Kuhn, Thomas. 1970. *Structure of Scientific Revolutions.* 2nd ed. Chicago: University of Chicago Press.

LaChance Adams, Sarah. 2011. "The Ethics of Ambivalence: Maternity, Intersubjectivity, and Ethics in Levinas, Merleau-Ponty, and Beauvoir." PhD diss., University of Oregon.

Lakoff, George. 1987. *Women, Fire, and Dangerous Things: What Categories Reveal about the Mind.* Chicago: University of Chicago Press.

———. 1992. *Moral Politics: What Conservatives Know that Liberals Don't.* Chicago: University of Chicago Press.

———. 2004. *Don't Think of an Elephant: Know Your Values and Frame the Debate.* White River Junction, VT: Chelsea Green Publishing.

———. 2006. *Whose Freedom?: The Battle over America's Most Important Idea.* New York: Farrar, Straus and Giroux.

———. 2008. *The Political Mind: Why You Can't Understand 21st-Century Politics with an 18th-Century Brain.* New York: Viking.

Lakoff, George, and Mark Johnson. 1980. *Metaphors We Live By.* Chicago: University of Chicago Press.

———. 1999. *Philosophy in the Flesh: The Embodied Mind and Its Challenge to Western Thought.* New York: Basic Books.

Lakoff, George, and Rafael Núñez. 2000. *Where Mathematics Comes From: How the Embodied Mind Brings Mathematics into Being.* New York: Basic Books.

Langacker, Ronald. 1987/1991. *Foundations of Cognitive Grammar.* 2 vols. Stanford, CA: Stanford University Press.

LeDoux, Joseph. 2002. *Synaptic Self: How Our Brains Become Who We Are.* New York: Viking.

Levinas, Emmanuel. 1969. *Totality and Infinity: An Essay on Exteriority.* Translated by Alphonso Lingis. Pittsburgh: Duquesne University.

Lewis, Clive S. 1952/2001. *Mere Christianity.* New York: HarperCollins.

Louden, Robert. 2000. *Kant's Impure Ethics: From Rational Beings to Human Beings.* Oxford: Oxford University Press.

Luu, Phan, and Don Tucker. 2003. "Self-Regulation by the Medial Frontal Cortex: Limbic Representation of Motive Set-Points." In *Consciousness, Emotional Self-Regulation and the Brain,* edited by M. Beauregard, 123–61. Amsterdam: John Benjamins.

Lyotard, Jean François. 1979. *The Postmodern Condition: A Report on Knowledge.* Translated by G. Bennington and B. Massumi. In *Theory and History of Literature.* Vol. 10. Minneapolis: University of Minnesota Press.

MacIntyre, Alasdair. 1984. *After Virtue.* 2nd ed. Notre Dame, IN: University of Notre Dame Press.

Mahlmann, Matthias, and John Mikhail. 2005. "Cognitive Science, Ethics, and Law." *Epistemology and Ontology*, 95–102.

Mallon, Ron. 2008. "Reviving Rawls's Linguistic Analogy Inside and Out." In *Moral Psychology.* Vol. 2: *The Cognitive Science of Morality, Intuition, and Diversity*, edited by W. Sinnott-Armstrong, 145–55. Cambridge, MA: MIT Press.

McCauley, Robert. 2011. *Why Religion Is Natural and Science Is Not.* Oxford: Oxford University Press.

Mernissi, Fatima. 1994. *Dreams of Trespass: Tales of a Harem Childhood.* Reading, MA: Addison-Wesley.

Mikhail, John. 2000. "Rawls' Linguistic Analogy: A Study of the 'Generative Grammar' Model of Moral Theory Described by John Rawls in 'A Theory of Justice.'" PhD diss., Cornell University.

———. 2007. "Universal Moral Grammar: Theory, Evidence and the Future." *TRENDS in Cognitive Science* 11 (4): 143–52.

———. 2008. "The Poverty of the Moral Stimulus." In *Moral Psychology.* Vol. 1: *The Evolution of Morality: Adaptations and Innateness*, edited by Walter Sinnott-Armstrong, 353–59. Cambridge, MA: MIT Press.

Moore, G. E. 1903/1968. *Principia Ethica.* Cambridge: Cambridge University Press.

Noddings, Nell. 1984. *Caring: A Feminist Approach to Ethics and Moral Education.* Berkeley: University of California Press.

Nussbaum, Martha. 1986. *The Fragility of Goodness: Luck and Ethics in Greek Tragedy and Philosophy.* Cambridge: Cambridge University Press.

———. 2000. *Women and Human Development: The Capabilities Approach.* Cambridge: Cambridge University Press.

Pappas, Gregory. 2008. *John Dewey's Ethics: Democracy as Experience.* Bloomington: Indiana University Press.

Pincoffs, Edmund. 1986. *Quandaries and Virtues: Against Reductionism in Ethics.* Lawrence: University Press of Kansas.

Prinz, Jesse. 2004. *Gut Reactions: A Perceptual Theory of Emotion.* Oxford: Oxford University Press.

———. 2008. "Is Morality Innate?" In *Moral Psychology.* Vol. 1: *The Evolution of Morality: Adaptations and Innateness*, edited by Walter Sinnott-Armstrong, 367–406. Cambridge, MA: MIT Press.

Putnam, Hilary. 1981. *Reason, Truth and History.* Cambridge: Cambridge University Press.

———. 1987. *The Many Faces of Realism.* LaSalle, IL: Open Court.

———. 2004. *Ethics without Ontology.* Cambridge, MA: Harvard University Press.

Quine, Willard V. O. 1960. *Word and Object.* Cambridge, MA: MIT Press.

Rawls, John. 1971. *A Theory of Justice*. Cambridge, MA: Harvard University Press.
———. 1980. "Kantian Constructivism in Moral Theory." *Journal of Philosophy* 77(9): 515–72.
Ricoeur, Paul. 1992. *Oneself as Another*. Translated by Kathleen Blamey. Chicago: University of Chicago Press.
Rizzolatti, Giacomo, and Laila Craighero. 2004. "The Mirror-Neuron System." *Annual Review of Neuroscience* 27: 169–92.
Rorty, Richard. 1979. *Philosophy and the Mirror of Nature*. Princeton, NJ: Princeton University Press.
———. 1982. *Consequences of Pragmatism*. Minneapolis: University of Minnesota Press.
———. 1989. *Contingency, Irony, and Solidarity*. Cambridge: Cambridge University Press.
Rosch, Eleanor. 1973. "Natural Categories." *Cognitive Psychology* 4: 328–50.
———. 1977. "Human Categorization." In *Studies in Cross-Cultural Psychology*, edited by N. Warren. London: Academic Press.
Royce, Josiah. 1908. *The Philosophy of Loyalty*. New York: Macmillan.
Rozin, Paul, Jonathan Haidt, and Clark McCauley. 2008. "Disgust." In *Handbook of Emotions*, edited by M. Lewis and J. Haviland, 575–94. New York: Guilford.
Sandel, Michael. 1998. *Liberalism and the Limits of Justice*. Cambridge: Cambridge University Press.
Sartre, Jean-Paul. 1946/1956. "Existentialism Is a Humanism." Translated by Philip Mairet. In *Existentialism from Dostoevsky to Sartre*, edited by Walter Kaufmann. New York: Meridian Books.
Schulkin, Jay. 2011. *Adaptation and Well-Being: Social Allostasis*. Cambridge: Cambridge University Press.
Seligman, Martin, and Mihaly Chikszentmihalyi. 2000. "Positive Psychology: An Introduction." *American Psychologist* 55 (1): 5–14.
Shafer-Landau, Russ. 2003. *Moral Realism: A Defence*. Oxford: Oxford University Press.
Sheets-Johnstone, Maxine. 2008. *The Roots of Morality*. University Park: Pennsylvania State University Press.
Short, Elizabeth, Damien Riggs, Amaryll Perlesz, Rhonda Brown, and Graeme Kane. 2007. *Lesbian, Gay, Bisexual, and Transgender (LGBT) Parented Families: A Literature Review Prepared for the Australian Psychological Society*. Melbourne: Australian Psychological Society.
Shusterman, Richard. 2000. *Pragmatist Aesthetics: Living Beauty, Rethinking Art*. Lanham, MD: Rowman & Littlefield.
Shweder, R. A., and E. Bourne. 1984. "Does the Concept of the Person Vary Cross-Culturally?" In *Cultural Theory*, edited by R. Shweder and R. LeVine, 158–99. Cambridge: Cambridge University Press.
Shweder, Richard, Manamohan Mahapatra, and Joan Miller. 1987. "Culture and Moral Development." In *The Emergence of Morality in Young Children*, edited by J. Kagan and S. Lamb, 1–83. Chicago: University of Chicago Press.

Shweder, Richard, Nancy Much, Manamohan Mahapatra, and Lawrence Park. 1997. "The 'Big Three' of Morality (Autonomy, Community, and Divinity), and the 'Big Three' Explanations of Suffering." In *Morality and Health*, edited by A. Brandt and P. Rozin, 119–69. New York: Routledge.

Singer, Peter. 1981. *The Expanding Circle*. New York: Farrar, Straus, Giroux.

Sripada, Chandra. 2008. "Nativism and Moral Psychology: Three Models of the Innate Structure that Shapes the Content of Moral Norms." In *Moral Psychology*. Vol. 1: *The Evolution of Morality: Adaptations and Innateness*, edited by Walter Sinnott-Armstrong, 319–43. Cambridge, MA: MIT Press.

Stern, Daniel. 1985. *The Interpersonal World of the Infant: A View from Psychoanalysis and Psychological Development*. New York: Basic Books.

Stuhr, John. 1997. *Genealogical Pragmatism: Philosophy, Experience, and Community*. Albany: State University of New York Press.

Sullivan, Maureen. 2004. *The Family of Woman: Lesbian Mothers, Their Children, and the Undoing of Gender*. Berkeley: University of California Press.

Thagard, Paul. 2010. *The Brain and the Meaning of Life*. Princeton, NJ: Princeton University Press.

Trevarthen, Colwyn. 1979. "Communication and Cooperation in Early Infancy: A Description of Primary Intersubjectivity." In *Before Speech: The Beginning of Interpersonal Communication*, edited by Martha Bullowa, 321–48. Cambridge: Cambridge University Press.

Tucker, Don. 2007. *Mind from Body: Experience from Neural Structure*. Oxford: Oxford University Press.

Turiel, Elliot. 2002. *The Culture of Morality*. Cambridge: Cambridge University Press.

Warnock, G. J. 1967. *Contemporary Moral Philosophy*. Oxford: Oxford University Press.

Williams, Bernard. 1985. *Ethics and the Limits of Philosophy*. Cambridge, MA: Harvard University Press.

Wilson, Edward. 1978. *On Human Nature*. Cambridge, MA: Harvard University Press.

Winter, Steven. 1990. "*Bull Durham* and the Uses of Theory." *Stanford Law Review* 42 (3): 639–93.

———. 2001. *A Clearing in the Forest: Law, Life, and Mind*. Chicago: University of Chicago Press.

Wittgenstein, Ludwig. 1953. *Philosophical Investigations*. Translated by G. E. M. Anscombe. Oxford: Blackwell.

Wright, H. G. 2007. *Means, Ends and Medical Care*. Dordrecht: Springer.

Index

absolutism. *See* fundamentalism; moral fundamentalism
abstraction, 94, 96, 98, 99–100, 107, 163, 232nn2–3
  within moral fundamentalism, 163
  of objects/events (Dewey), 232n2
  Psychologist's Fallacy (James), 232n3
  varieties of, 100
  See also reductionism
action, 103, 105, 108, 115, 215–16
  moral judgments directed at, 147
  unity of self and action (Dewey), 215–16
  See also motivation
activism, 106
Adams, Sarah LaChance, 60, 230n6
aesthetic judgment. *See* judgment: aesthetic
aesthetic sensitivity, 68
  as *eudaemonistic* virtue, 68
  as ground of meaningful experience, 68
  as ground of morally significant sociality, 68, 117
  as "moral artistry," 117
  and reasonableness, 118
  See also art; experience: aesthetic/felt quality of
affordances, 95
after-the-fact justification. *See* rationalization
aggression, 153–54

allostasis, 55, 77, 78, 80, 83, 85, 93, 202, 231n5
  collective/social, 85
  definition of, 55
  as dynamic process of life maintenance, 55
  as ground of subsidiary biological values, 55–56
  relevance to emotional response patterns, 78
  as unification of experience, 202
  See also homeostasis
altruism, 125. *See also* self-interest
animals. *See* nonhuman animals
argument. *See* ethical argument
Aristotle, 116, 207–8
art, 97–98, 117–18, 213, 220, 233n4, 237n12
  artworks as events, 233n4
  literature, 213, 237n12
  See also aesthetic sensitivity; judgment: aesthetic; moral artistry
attachment, 59
authority, 65, 70, 71, 167
  acknowledgment of, as social virtue, 65, 71, 167
  children's beliefs regarding figures of, 36
  of ethical principles, 142; children's perceptions of, 36
  of God, as ground of moral systems, 166
  metaphorical structures of, 71, 179; within moral theories, 179

authority (*cont.*)
    of sacred text, as problematic, 11, 12, 210
    of universal reason, as ground of moral systems, 166
authority/subversion foundation of moral systems (Haidt), 70, 71, 179
autonomy, 30, 73–74, 83
    as basic dimension of moral systems, 161
    defined as positive freedom, 6
    distinguished from obedience to authority, 65
    (rational) as origin of moral principles, 6
"autonomy of the ethical" misconception, 30, 34–46, 182
    grounded in myth of uniquely "moral" experience, 34–40
    grounded in myth of uniquely "moral" judgment, 40–46
    historical origins of, 34–46
    as vacuous concept, 182
    See also experience: distinctly moral; Kant, Immanuel: Kantian moral theory

Baier, Annette, 116, 233n2
Barsalou, Lawrence, 107
Bechtel, William, 21–22
Bergen, Ben, 108–9
Bernard, Claude, 54
biological-cultural matrix, 4. *See also* organism-environment interaction
body, 107–9
    as origin of concepts, 107–8
    as origin of emotions, 8, 98, 102 (*see also* emotions)
    See also organism-environment interaction; reason: embodied
brain, human 59, 62, 68, 69, 75, 76, 80–82, 84, 109, 142, 144–45, 148–51, 155, 234n3
    and behavior, 148–49
    damage, 80–82, 231n8
    hormone-brain interactions, 59
    and the imagination, 109 (*see also* imagination)
    interconnectedness of, 148–51
    mirror neurons, 61–62, 109, 147
    "modules," 144, 145, 152
    and moral grammar arguments, 155
    neural plasticity, 68, 69
    prefrontal cortex, 84
    See also cognitive science

"Capabilities" approach, 56, 60
care, 56–57, 58–61, 65, 70, 229n5
care ethics, 59–61, 230n6
care/harm foundation of moral systems (Haidt), 70, 179
categories, 170, 171, 177, 213
    classically structured (Lakoff), 170, 171, 177, 179; defined, 1
    of personhood, 170–80 (*see also* moral personhood)
    radial-category theory (Lakoff), 177, 213
    See also concepts
character, 64, 92, 105, 196–98
    self-in-process, 197–98, 215, 218
    See also habits; personal identity; virtues
children. *See* developmental psychology; parenting
choice. *See* decision-making; Dewey, John: choice
Chomsky, Noam, 140–41, 143–46, 161
Christianity. *See* religion
Churchland, Patricia, 58–59, 62, 148–49
Churchland, Paul, 32, 143
circularity (within moral theory), 11
civic-mindedness, 65, 230n8
cognition, 51, 73–88, 89–111, 149
    cognitive capacities, 51, 235n6; necessary for moral personhood (Kant), 174
    and interconnectedness of neural systems, 148–49
    nonconscious/pre-reflective, 73–88, 124, 142, 195 (*see also* moral deliberation: "intuitive track")
    See also brain, human; decision-making; imagination; moral deliberation
cognitive-conative-affective simulation, 107–11
cognitive science, 23, 73–74, 107–9, 130, 138, 144, 148–51, 196
    cognitive linguistics, 25, 108–9, 142, 144; neural theory of language, 145–46; relevance to understanding moral deliberation, 142
    cognitive neuroscience, 138; relevance to moral theory, 138, 148–51, 196, 220
    See also brain, human; theory of mind
conatus, 79

concepts, 25, 95, 101, 107–8, 169–80, 190, 196, 204–5, 213, 215, 233n8
conceptual systems, 107–8, 173, 196
  as context-dependent, 173, 177, 178, 204–5, 213
  fixed, as problematic, 172, 173, 175–77, 178, 180–81, 205, 213, 215
  literalist conception of, within Moral Law folk theory, 169, 170–80, 177, 215; as unsound, 177, 179–80
  mathematical, 173
  moral concepts, 169–80, 215 (*see also* moral personhood: concepts of)
  relevance to ethical naturalism, 23–27, 196
  See also categories; metaphors
conflict, 30, 48–49, 79, 82, 92, 94, 95–100, 102, 103–7, 110, 113–14, 126, 196
  resolution, 82, 102, 105, 110, 113–14, 118, 214; as goal of moral inquiry, 214
  See also problem
conscientiousness, 93, 106, 213–16, 219
  opposed to dogmatism, 216
  as principal virtue of moral inquiry, 219
  See also critical perspective
consequences, unpredictability of, 202–3
consequentialism, 45, 156, 189
  and calculation of magnitude of harms, 156
  dynamic consequentialism (Kitcher), 189
  as potentially fundamentalist, 189
convention (social/cultural), 35–38, 105–7
  vs. morality, 37, 40, 155
courage, 64–65
Craig, Joseph, 70, 234n1
creativity, 68–69
critical perspective, 32, 87, 89–90, 91–93, 94, 103, 119–24, 127, 136
cultural relativism, x, 38, 59, 63, 69, 105–7
  See also moral relativism; values: culturally/historically situated
culture, 3, 4, 105–7
  cross-cultural studies, value of, 195–96
  relationship to nature, 3, 4, 5–53, 208, 218
  See also dualisms: nature/culture; sociality; values: cross-cultural; values: culturally/historically situated
Cushman, Fiery, 152, 158

Damasio, Antonio, 54–56, 77–82, 84, 85–87, 110, 150, 231nn6–8
Darwin, Charles, 124. *See also* evolution
decency, x–xi
decision-making, 81, 82, 89, 110, 114, 134, 201. *See also* cognition; conflict; Dewey, John: choice; inquiry; moral deliberation; problem: problem-solving
deliberation. *See* moral deliberation
desire, 82, 115–19
  for moral order, 164–69
  See also dualisms: reason/desire
developmental psychology, 35–36, 40, 51, 57–60, 131, 161–62. *See also* organism-environment interaction
Dewey, John, 24–25, 30, 34, 39, 52, 77, 79, 82, 86, 87–88, 91–102, 103, 114, 115, 116–17, 118–19, 120, 134, 139, 153–54, 162, 184–85, 186–87, 188, 190–91, 197–202, 214–18, 219, 220, 227n3, 232nn1–2, 233n1, 233n4, 233n6, 236nn3–4
  body-mind, 24
  choice, 114, 117, 197
  critical reflection, 87–88
  ends-in-view, 52, 108, 214, 228n3
  experience, 39, 66, 87, 94–96, 98
  fact, 186–87
  ideas, 186
  inquiry, 93–94, 99, 100, 130, 186–87, 190–91, 197–98, 200, 216 (*see also* Dewey, John: logic)
  instinct, 139, 153–55
  logic, 184, 186, 233n1 (*see also* Dewey, John: inquiry)
  moral deliberation, 93–94, 161, 203–13, 216, 228n3; example of, 203–13
  pervasive unifying quality, 94–102, 233n4
  pragmatist categorical imperative, 202
  principles, 188, 191
  problem-solving, 30, 79, 86, 200, 214, 216, 228n3
  reasonableness, 82–83, 114, 116–17, 118–19, 134
  selective interest, 39
  self-evidence, 184–85
  self-in-process, 197–98, 215–16, 218
  valuation (reflective), 91, 93, 197, 212, 214, 216, 219
  valuing (intuitive), 52, 91, 93, 197, 219
  willing, 25

INDEX

dignity, 123, 229n10
dilemmas. *See* moral dilemmas
disgust reactions, 71, 158
    as component of moral faculty (Hauser) 147
divine command theory. *See* religion: Judeo-Christian
Donagan, Alan, 113, 171, 172, 178, 181, 215, 229n11
dualisms
    fact/value, 44
    mind/body, 8, 16, 24, 109, 148–49
    moral fundamentalism/moral relativism, 194–95
    nature/culture, 3, 51, 131–32, 208, 218 (*see also* culture: relationship to nature)
    reason/desire, 82, 115–19
    reason/emotion, 74–75, 76, 81–84, 91, 114, 115–16, 150
    subjective/objective, 98, 99
Dweyer, Susan, 144, 155–57, 234n4

Edel, Abraham, 3
Edelman, Gerald, 149
egoism. *See* self-interest
emotions, 24, 61, 74, 75, 76, 77–85, 98, 102, 109–11, 114, 115–16, 193, 231nn6–8
    allostasis and emotional response patterns, 78
    as basis for sociality, 80–82
    consciousness of, 78–79
    empathetic (*see* empathy)
    as ground of morality (Damasio), 231n8
    gut feelings, 81
    See also dualisms: reason/emotion; reason: and emotions
empathy, 59–62, 80, 109, 120, 230n7
    as component of moral faculty (Hauser), 147
ends, 107, 108, 110, 134–35, 215, 227n1
    ends-in-view, 52, 108, 214, 228n3
    internal/external, 227n1
    See also good: end
Enlightenment, 24, 41, 74, 76, 151, 153
ethical argument, 118, 126, 135
    within non-naturalistic ethical theories, 226n10
ethical naturalism, 3, 14, 20–27, 28, 34, 45, 47–48, 52–53, 61, 86, 129–33, 161, 181, 195–96
    as alternative to "moral faculty" view, 137–62
    as alternative to moral fundamentalism, 14, 86, 129–33, 181, 196, 200
    defined, 3
    descriptive dimension of, 129–32
    distinguishing characteristics of, 196
    need for, 24–26
    normative dimension of, 132–36
    pragmatic naturalism, 125–29, 131
    reasonableness within, 15, 82–83, 90, 112–36, 193 (*see esp.* 113–14, 118, 134) (*see also* Dewey, John: reasonableness)
    rejected by Kant, 45–47
    role of science within, 20–23, 124, 129, 137–39, 181, 192–93, 203, 219–20, 236n2
    See also moral deliberation; organism-environment interaction; values: "maturationally natural"
"ethical project" (Kitcher), 125–29
ethics. *See* ethical naturalism; Kant, Immanuel: Kantian moral theory; moral theories; values
eudaemonia. *See* flourishing; happiness
Euthyphro. *See* Plato
evolution, 54, 56, 58–59, 69, 70–72, 79, 84, 91, 102, 124, 148, 159, 161, 208, 218
    evolutionary psychology, 137; relevance to moral theory, 137–38
    as origin of cross-cultural values, 138–39, 161
    See also genes; organism-environment interaction
existentialism, 166
experience, 29, 33–40, 94–98, 101–2, 159–60, 194, 234n4
    aesthetic/felt quality of, 97–99, 101–2, 117–18, 121
    distinctly moral, 29, 33–40, 159–60 (*see also* folk theories: Experiential Kinds; judgment: distinctly moral)
    as multi-dimensional, 38–40, 48, 94–95, 96–98, 99, 100, 151, 160, 192, 234n4
    pervasive unifying quality of, 94–102
    See also Dewey, John: experience

facts, 180–81, 185–87
    definition of, within moral realism, 185–86

INDEX

Dewey's conception of, 186–87
relevance to moral reasoning (Hume), 116
See also moral facts; moral realism
fact/value dichotomy. *See* dualisms: fact/value
faculty psychology, 24–25, 41–42, 74, 84, 114, 116, 137–62 (*see esp.* 146–48, 151)
as empirically suspect, 144–45, 146, 148–51
grammatical faculty, 142–46; typical arguments supporting, 155–57 (*see also* instinct: language)
moral faculty, 137–162 (*see esp.* 146–48), 234n2, 234n4; arguments supporting, 141, 146–49, 155–57; as bulwark against moral relativism, 234n4; children's purported possession of, 155; components of (Hauser), 147–48; Damasio's rejection of, 150–51; as empirically suspect, 146, 148–51, 155–57; Hauser's defense of, 141, 146–49; Mallon's response to, 234n3; morality without, 159–62; need for the rejection of, 139, 151, 155, 162
See also instinct: moral instinct; linguistics: Chomskyan
fairness, 70. *See also* injustice
fairness/cheating foundation of moral systems (Haidt), 70, 179
falsifiability
aversion to, 163
within the sciences, 23, 120
feelings. *See* emotions
Feldman, Jerome, 144–45
Fesmire, Steven, 117–18
Feyerabend, Paul, 120–21
Flanagan, Owen, 53–54, 56, 80, 130, 132–35, 152–53, 161, 225, 227n15
flexibility. *See* open-mindedness
flourishing, 53–56, 59, 61, 65–68, 77–80, 83, 86–87, 102, 128
social dimensions (*see* sociality)
See also care; care ethics; happiness; organism-environment interaction
folk theories
Experiential Kinds, 41, 159–60, 194
Moral Law, 16, 24–25, 102, 113, 124, 169–70, 177, 215, 217; conception

of self within, 215; summarized, 160–70
Foucault, Michel, 105–6
Frankl, Viktor, ix–x
freedom, 122, 201–2
pragmatic notion of, 201–2
See also free will
free will, 6, 8, 25, 73, 124, 170, 227n15
radical (within Moral Law folk theory), 170
religious views concerning, 8
See also freedom
Freud, Sigmund, 124, 165–66
fulfillment. *See* meaning: human need for
fundamentalism, 68, 119, 122, 124, 163–80, 211. *See also* moral fundamentalism

Galilei, Galileo, 144
Gallagher, Shaun, 57–58
Geertz, Clifford, 164–65
gender, 60
genes, 148
Gibson, J. J., 95
Gilligan, Carol, 59–60
good, 79, 136, 216
"better/worse," as alternative to good/bad, 220
end, 31, 52, 107, 198, 214 (*see also* ends)
intrinsic vs. extrinsic, 50, 52
static conception of, within utilitarianism, 189
goodness
as non-natural property 19–20, 135, 182, 183; Warnock's argument against, 183
of universe (C. S. Lewis), 167
growth, 196–202
as a basis for morality, 198–202, 212
of meaning (of a situation), 200, 202, 212
as overcoming of present self-identity, 199

habits, xii, 25, 68, 91, 94, 103, 105, 122, 196–97, 220
insufficient to resolve novel conflicts, 197, 199, 220
willing as complex interpretation of, 25
Haidt, Jonathan, 38, 65, 69–72, 75, 76, 89–90, 153, 161, 179, 226n8, 230n2, 230n10, 231n4, 234n1, 237n10

Haidt, Jonathan (cont.)
  foundations of moral systems, 70–72, 170, 237n10
  social intuitionist model of moral assessment, 75–76
happiness, 66–68
  in the afterlife, 166
  See also flourishing; positive psychology
harm, 36, 39–40, 70, 119, 171
  calculation of magnitude of, 156
  direct/indirect, beliefs concerning, 156
  general moral principles against, 158
  prohibitions against based on respect for persons, 171
Harman, Gilbert, 157–58
Hauser, Marc, 34, 40, 75, 76, 89–90, 138–54, 155–57, 158, 161, 228n4, 230n2, 231n3, 234nn1–4, 235n6
  components of the moral faculty, 147–48
  linguistic analogy, 140, 141–43, 234n4; strong version, 143
  See also faculty psychology: moral faculty, Hauser's defense of
health, 39, 214. See also flourishing
Hickman, Larry, 228n3
Hill, Thomas, Jr., 174, 175
Hinde, Robert, 59, 80, 131–32
Hoffman, Martin, 61
homeostasis, 54–55, 77–78, 80, 85–88, 93, 110, 231n5
  collective/social, 85–86
  definition of, 55
  inclusive of allostasis, 55
  See also allostasis
honesty. See truthfulness
Huebner, Bryce, 144, 155–57, 234n4
human finitude, xi, 219, 235n1
  evasion of, 219
human nature/human situation, 3, 80, 124, 126, 149, 153, 154, 202, 208, 217–19, 225n1, 236n2
  as both natural and cultural, 3, 208
  and instincts, 153, 154
  Kantian views regarding, 44
  Law of Human Nature (C. S. Lewis), 166–67
  modest notion of (within ethical naturalism), 15, 53, 124, 130–32, 149, 217–18
  religious beliefs regarding, 5
  See also nonhuman animals; organism-environment interaction; sociality
Hume, David, 61, 75, 115–16, 134, 233n2
humility, 63–64
humor, 220
Hutto, Daniel, 57–58

ideals, 128, 193, 219. See also moral principles; values
identity. See personal identity
imagination, xi, 74, 90–91, 93–94, 105, 107–11, 117, 220. See also moral artistry; moral deliberation: as imaginative process
injustice, 106. See also fairness
inquiry, 31–33, 93–94, 99, 100, 110, 119, 120, 125, 190–91, 194–95, 213
  forms of, 31
  initial stage of, 103–7, 195, 199–200 (see also problem: defining: as first stage of moral inquiry)
  as lifelong process, 191, 220
  logic of, 100, 184, 186, 233n1
  moral insight as the result of, 213
  See also Dewey, John: inquiry; Dewey, John: logic; moral deliberation; problem: problem-solving
instinct, 84, 138, 153–55, 155–57, 158
  for aggression, 153–54
  as ground of universal human behaviors, 153–55; Dewey's rejection of, 153–54
  language, 140 (see also faculty psychology: grammatical faculty; linguistics: Chomskyan)
  moral instinct, 138, 139, 140, 153–55, 234n1; Dewey's rejection of, 139
  See also faculty psychology: moral faculty; moral grammar
integrity, 64
intelligence, 68, 80–81, 119, 202, 218, 220, 230n9
intentionality, 57, 58
intersubjectivity, 57–58, 60, 63. See also sociality
intuition, 73–88, 89–90, 92, 116, 195, 231n4
  moral intuition 75, 81, 82, 101, 102, 116
  See also faculty psychology: moral faculty; instinct: moral instinct; moral deliberation; moral grammar
is-ought fallacy. See naturalistic fallacy

INDEX

James, William, 232n3
Johnson, Mark, 12, 16, 71, 169–70, 234n1, 236n4
judgment, 2, 28–30, 40–46, 75, 76, 77, 89–90
  aesthetic, 100–101, 118, 161, 195; similarity to moral judgment, 68, 117, 118, 161, 195 (*see also* aesthetic sensitivity; art; experience: aesthetic/felt quality of; moral artistry)
  distinctly moral, 29, 35, 37, 40–46, 124, 151, 159–60 (*see also* folk theories: Experiential Kinds)
  evaluative, with radial-category theory, 177
  felt quality (*see* experience: aesthetic/felt quality of)
  relationship between aesthetic and moral, 68, 101–2 (*see also* aesthetic sensitivity; moral artistry)
  See also Kant, Immanuel: Kantian moral theory, judgment; moral deliberation

Kant, Immanuel
  "Critical Philosophy," 42
  Kantian moral theory, 6–13, 18–19, 41–47, 50, 61, 64, 83, 115, 158, 166, 171, 174, 182, 215, 226nn5–6, 227n12, 228nn6–12; hypothetical/categorical imperatives, 19, 44–45, 215 (*see also* Dewey, John: pragmatist categorical imperative); judgment (definition of), 42, 228nn6–7; relationship to Judeo-Christian ethical thought, 6, 13, 18, 43, 46; respect for persons, 171, 172, 229n10 (*see also* respect for persons); views on sexual permissibility, 9 (*see also* pure practical reason)
  "moral argument" for God's existence, 164
Kierkegaard, Søren, 237n11
Kitcher, Philip, 125–29, 189
knowledge
  absolute moral, 169–80 (*see also* moral fundamentalism)
  foundational, 26
  moral, as technology, 227n3
  scientific, 120, 181 (*see also* ethical naturalism: role of science within; science)
  varieties of relevant to judging, 195–96

Kohlberg, Lawrence, 60
Koopman, Colin, 101, 105–6

Lakoff, George, 71, 171, 230n10
language, 108–9, 140, 141, 161
  acquisition, neural theory of language view of, 145–46, 161
  embodied construction grammar, 145–46, 161
  linguistic analogy, 140–43, 155, 158, 161, 234n4
  linguistic competence and moral competence compared, 143, 158, 161–62
  See also cognitive science: cognitive linguistics; linguistics
laws, 65
Levinas, Emmanuel, 226n11
Lewis, C. S., 73, 166–69
linguistics, 140
  Chomskyan, 140–41, 143–46, 161; "poverty of the stimulus" argument, 140–41
  cognitive linguistics, 25, 108–9, 142, 144; neural theory of language, 145–46; relevance to understanding moral deliberation, 142
  See also language; moral grammar
logic, 100, 113, 174
  distinguished from rationality, 174
  See also Dewey, John: logic
Louden, Robert, 227n13, 228n9, 229n11
loyalty, 65, 70, 220
loyalty/betrayal foundation of moral systems (Haidt), 70, 179
luck, 220, 236n7
Luu, Phan, 55

MacIntyre, Alasdair, 31–33, 63, 69, 127, 172, 184, 213–14, 227n1
  internal/external ends, 227n1
  practice/*praxis*, 31–32, 63, 184, 213–14, 227n1
Mallon, Ron, 139, 234n3
marriage, 9, 10, 203–10, 212, 213, 236nn5–6
  gay marriage, 203–10, 212
  radial category structure of, 213
McCauley, Robert, 51, 143, 235n1
meaning, 1, 25, 54, 86, 95, 99, 108–9, 144, 150, 184–85, 200, 202, 236n4
  Dewey's pragmatist theory of, 236n4

255

INDEX

meaning (*cont.*)
    embedded in social context, 35, 95, 184–85
    human need for, 54, 66–68
    increase in as moral success, 200, 202
    literalist theory of, 25
    moral significance, 40, 104
    semantics, 144
medicine, 120
memory, 80–81, 107, 147
metaphors
    bonding, 179
    elephant and rider (Haidt), 76, 90
    for freedom, 179
    "moral accounting," 179
    moral purity, 71, 179
    of moral thinking, 193–96; arbitrary creation (values ex nihilo view), 194; creative transformation (pragmatist), 194–95; discovery (objectivist view), 194
    for sex, 7–8, 10, 225nn1–2
    strict parent/nurturant parent (authority), 71, 179, 230n10
    up/down, 225n4
metaphysics, 193–96, 236n8
    objectivist, 194
    pragmatist, 194
Mikhail, John, 138
mind, 24
    embodied, 24
    See also brain, human; dualisms: mind/body; mirror neurons
mirror neurons, 61–62, 109, 147
"mirror stage," 57
Moore, G. E., 19–20, 34, 135, 181–82, 183
"moral accounting" metaphor, 179
moral argument. *See* ethical argument
moral artistry, 117–18, 194–95
moral confusion. *See* moral fundamentalism: personal suspicions regarding
moral deliberation, xi, 1, 2, 26–27, 48, 73–88, 89–111, 112–36, 141–42, 160, 197–98, 203–13, 214, 219, 220, 227n3
    critical dimension of, 119–24
    Deweyan, as a form of reflective valuation, 93–111; example of, 203–13, 214, 216, 219 (*see also* Dewey, John: inquiry)
    fast (intuitive) vs. slow (reflective), 2, 75–76, 81–83, 87, 89–90, 141–42, 193, 230n2
    as a form of problem-solving (or inquiry), xi, 2, 28–34, 48, 90–91, 93–94, 98–99, 103–7, 134, 149, 160, 187, 192, 194–95, 199–200; example of, 203–13, 214, 216, 219, 227n3
    as imaginative process, 93–94, 105, 107–11, 121, 193, 194, 220 (*see also* imagination)
    "intuitive track," 73–88, 89–90, 92, 116, 195, 231n4 (*see also* cognition: nonconscious/pre-reflective; intuition: moral intuition)
    as means to character development, 197
    open-endedness of, 25
    and qualitative unity of a situation, 96–102, 103
    "reflective track," 87, 116 (*see also* critical perspective; Dewey, John: critical reflection)
    as *sui generis*, 33, 37, 45
    "third track," 89–91, 193, 194–95, 197, 219 (*see also* moral deliberation: as a form of problem-solving)
    See also critical perspective; judgment: distinctly moral; valuation
moral dilemmas, 156–58
moral facts, 163, 164, 166, 169, 180–83, 185–87, 190. *See also* moral realism
moral fundamentalism, x–xi, 1, 2, 26–27, 73–74, 124, 133, 163–91 (*see esp.* 163–64), 200, 202, 209–11, 219
    attractiveness of, 163–69
    based on misconception of mind/thought, 169–80, 210
    as cognitively indefensible, xi, 4, 14, 190, 193, 210
    C. S. Lewis's argument for, 166–69
    definition of, 164
    fear of abandoning, 124, 219
    immorality of, 190–91, 192, 200, 202, 211, 219
    Kantian, 6, 42–47 (*see also* Kant, Immanuel: Kantian moral theory)
    moral instinct theories as a form of, 163 (*see also* faculty psychology: moral faculty; moral grammar)
    personal suspicions regarding, 5–15

INDEX

premised upon false dichotomy, 168–69
religious, 5, 121, 163, 164–67, 204–6, 209–11, 237n11 (*see also* religion)
as threat to ethics, 4, 14, 26, 99–100, 122, 180, 190–91 (*see also* moral fundamentalism: as immoral)
two main types of, 163
within utilitarianism, 189
See also dualisms: moral fundamentalism/moral relativism; theories of value, non-naturalistic
moral grammar, 138, 140–43, 151–52, 155–59, 234n2
arguments supporting, 155–57; and multiple conceptions of universality, 157–59
attributed to Chomskyan linguistics, 140–42
based on problematic analogy, 140–42, 158 (*see also* language: linguistic analogy)
definition of, 140
paucity of evidence for, 155–57
See also faculty psychology: moral faculty; instinct: moral instinct
Moral Imagination. *See* Johnson, Mark
morality (fit) for humans, 1–2, 124, 191, 192–93, 217–21
argument for, summarized, 192–93
defined, xii, 1–2
moral judgment. *See* judgment: distinctly moral; moral deliberation
moral law. *See* Kant, Immanuel: Kantian moral theory
Moral Law folk theory. *See* folk theories: Moral Law
moral obligation, 29, 118, 227n12. *See also* moral principles
moral order, 164–69
C. S. Lewis's argument for, 166–68
moral personhood, 123, 126–27, 170–80
and artificial intelligence, 178
concepts of, 170–80; outside the Judeo-Christian tradition, 178–79
and corporations, 178
distinguished from personality, 173
and the more-than human world, 177–78 (*see also* nonhuman animals)
problem of marginal cases, 175–77

prototypical instantiation of, 175
See also respect for persons
moral principles, 10, 76, 96, 100, 101, 113, 118, 121, 131, 166–69, 180–81, 186, 187–89, 205
as absolutes, x, 100, 113, 121, 131, 166–69, 201, 205, 215; within Moral Law folk theory, 169, 215 (*see also* Kant, Immanuel: Kantian moral theory; moral fundamentalism)
as hypotheses, 188 (*see also* moral deliberation: as a form of problem-solving)
non-foundationalist conception of, 187–89
problem of specificity, 10
self-evident (within moral realism), 180–81, 188 (*see also* moral facts; moral realism)
See also values
moral problem. *See* conflict; problem
moral progress
collective, 125–29, 135–36
personal (moral development), 197–202
moral purity. *See* purity; sacredness/sanctity
moral realism, 164, 180–87, 190
defined, 180
as a form of moral fundamentalism, 180–87, 190
non-naturalistic (Schafer-Landau), 182
reasons for popularity of, 164
See also moral facts; self-evidence
moral reasoning. *See* moral deliberation
moral relativism, 1, 13, 111, 119, 121, 168, 194–95, 234n4
American aversion to, 168
as challenge to ethical naturalism, 14, 111, 119; refuted, 195
moral fundamentalism as bulwark against, 168, 181, 234n4
See also cultural relativism; dualisms: moral fundamentalism/moral relativism; values: culturally situated
moral responsibility, 59, 73, 170, 201, 226n11
as component of moral faculty (Hauser), 147
forward-looking, 59
and free will, 6, 170, 201
within Moral Law folk theory, 170

INDEX

moral significance. *See* meaning: moral significance
moral theories/moral systems, 95, 96, 129–36
  abstraction within, 96
  basic dimensions of, 70–71, 161
  descriptive dimension of, 129–32
  metaphysics of, 193–96
  need for justificatory framework for, 11
  normative dimension of, 132–36
  See also care ethics; circularity; folk theories: Moral Law; Kant, Immanuel: Kantian moral theory; morality (fit) for humans; theories of value
mothering, 60. *See also* care; ethics of care; parenting
motivation, 75, 76, 83, 87, 180
  principles as sufficient for (Schafer-Landau), 180

natural
  as opposed to supernatural, 4, 217, 218
  unnaturalness, 207–8
naturalistic ethics. *See* ethical naturalism
naturalistic fallacy, 19, 34, 92–93. *See also* prejudice
nature. *See* culture: relationship to nature; ethical naturalism; organism-environment interaction; values: and the body
neural plasticity. *See* brain, human: neural plasticity
Newton, Isaac, 124
Nietzsche, Friedrich, 236n3
Noddings, Nel, 60
nonhuman animals, 73, 83–84, 86, 123, 126, 169, 174, 177–78, 217–18, 229n10
  distinguished from human beings via fixed concept of rationality, 174
  extension of moral consideration to, 177–78, 229n10, 230n1
nonnaturalism (ethical), 182. *See also* moral realism
normalcy, 229n1
Nussbaum, Martha, 56

obedience. *See* authority
open-mindedness/flexibility of thought
  as a component of proper moral inquiry, 189, 216, 220

conscientiousness characterized by, 216
  as a social virtue, 68
organism-environment interaction, 49–56, 70–71, 77–79, 83–84, 86, 97, 99, 108, 129, 131–32, 148, 192, 194, 217–18
  as basis of morality (fit) for humans, 217
  as source of personal identity, 217–18
"ought," 133, 134
  as form of rational imperatives, 215
  See also moral principles; values

pacifism. *See* war, moral permissibility of
Pappas, Gregory, 96
parenting, 59, 209
  "Nurturant Parent" model of morality, 71, 230n10
  same-sex, 209
  "Strict Family Father" model of morality, 71, 230n10
  See also care; care ethics; developmental psychology; mothering
patience, as component of moral faculty (Hauser) 148
personal identity, 4, 25, 105, 196–202, 215–16, 217–18, 220
  absolutist/static conception of, 215, 216; need for the abandonment of, 216
  and the biological-cultural matrix, 4, 217–18
  embodied, 217–18
  manifested through action, 215–16
  narrative sense of (Hauser), 147
  as ongoing reformation of the self (self-in-process), 197–202, 215, 218, 220
  See also character
personhood. *See* moral personhood
philosophy of science, 21, 120–21, 184
Pincoffs, Edmund, 69
Plato, 18, 20, 66, 219
pluralism, 119–20, 121, 124, 129
  as method of inquiry, 120
political theory, 63, 65, 128–29, 230n10. *See also* sociality: complex/political
positive psychology, 67. *See also* happiness
power, 105–6
practice/*praxis*, 31–32, 63, 184, 213–14, 227nn1–2
pragmatist ethics. *See* ethical naturalism; morality (fit) for humans
predispositions, 158–59
prejudice, 92–93, 105, 120, 122–23

258

principles. *See* moral principles
Prinz, Jesse, 159
problem, 92, 95–96, 103–7, 113–14, 126–27, 133, 203–12
  common, as source of cross-cultural values, 149
  context-dependent, 95–96, 203
  defining (as first stage of moral inquiry), 103–7, 195, 200
  problem-solving, 86, 92, 103–7, 108, 126, 134, 195–96, 199, 203–12, 227n3; crude, 227n3; Deweyan, illustrated, 203–12; general structure of, 30–31, 199; role of empirical knowledge, 32, 195–96, 202, 209, 211–12, 219–20 (*see also* conflict; decision-making; Dewey, John: inquiry, problem-solving; inquiry; moral deliberation: as a form of problem-solving)
  solutions, as tentative, 202
  See also conflict
pure practical reason, 18, 24, 42–45, 61, 83, 115, 227n12
  need for the abandonment of, 2, 24
purity, 71, 211, 237n10
  moral purity metaphor, 71, 179
  See also sacredness/sanctity
purposiveness (distinguished from essentialist teleology), 208

rationality. *See* reason
rationalization (after-the-fact justification), 2, 75, 76, 89–91, 142, 193, 219, 226n6, 226n8
  within philosophy, 226n6
Rawls, John, 13, 14, 122, 128–29, 140, 226n10, 230n3
realism. *See* moral realism
reason, 74–75, 81–84, 115–19, 173–76, 215
  embodied, 74–75
  and emotions, 76, 81–84, 91 (*see also* dualisms: reason/emotion)
  rationality, as ground of moral personhood, 173–77, 215; as problematic, 175–77
  transcendental, xii, 16, 112–13
  universal, within Moral Law folk theory, 169
  See also dualisms: mind/body; dualisms: reason/desire; dualisms: reason/emo-

tion; intelligence; moral deliberation; pure practical reason
reasonableness, 90, 112–36 (*see esp.* 113–14, 118, 134), 193. *See also* Dewey, John: reasonableness; ethical naturalism: reasonableness within
reductionism, 114, 122, 190
  of human behavior to instinct, 153–55
  within moral fundamentalism, as immoral, 190
  within the sciences, 154–55
  See also abstraction; moral fundamentalism
reflective equilibrium ("wide" vs. "narrow"), 122–23, 128, 135–36
religion, 106, 121, 124, 235n1
  Catholicism, 205
  definition of (Geertz), 164
  hermeneutical problems within, 11, 12, 205, 211, 226n7; biblical literalism, 209–11
  Judeo-Christian, 7, 43, 121, 171–72, 204–7, 209–11, 237n10; Isaac and Abraham story, 237n10; sectarian disagreements within, 172, 205
  Lutheranism, 5, 13
  "moral argument" for God's existence (Kant), 164
  original sin, 5
  prophecy, 106
  Protestantism, 205
  reasons for the existence of, 164–65
  as source of values, 5, 8–12, 64, 106, 166, 204–6, 209–11
  See also moral fundamentalism: sacredness/sanctity; soul
resolution. *See* conflict
respect for persons, 171, 172, 175, 176, 210, 229n10
revenge, 119
Rorty, Richard, 101, 236n1
Rosch, Eleanor, 25
Royce, Josiah, 65

sacredness/sanctity, 17, 70–71
  biological origins of, 71
  relationship to disgust, 71
  See also religion
sanctity/degradation foundation of moral systems (Haidt), 70, 71, 179, 237n10
Sandel, Michael, 63

Sartre, Jean-Paul, 166
Schafer-Landau, Russ, 180–83, 185, 187–88, 227n12
Schulkin, Jay, 55
science, 20–23, 120, 124
   moral science (Dewey), 236n2
   problematic understanding of, within ethical realism, 181
   reductionism within, 154–55
   relevance to ethical naturalism, 20–27, 181, 195–96, 219–20 (see also problem-solving: role of empirical knowledge)
   reproductive technologies, 207
   scientific explanation, 22
   social sciences, 196, 208, 209
   See also cognitive science; ethical naturalism: role of science within; evolution; philosophy of science
scientism, 120
self-criticism. See critical perspective
self-evidence, 183–85, 187, 188, 190
   as bogus concept, 183–85, 187, 188, 190
   Dewey's definition of, 184
   Schafer-Landau's definition of, 183
   See also moral facts; moral realism
selfhood. See character; personal identity
self-interest, 58–60, 65, 79–80, 86, 96
sensitivity, 118–19, 182, 201, 220
   as component of moral problem-solving, 201, 220
   and knowledge of the good, 182
   opposed to moral abstraction/reductionism, 118
   See also aesthetic sensitivity
sentiments, 61, 75, 100, 115–16. See also emotions; Hume, David
sex
   Kant's views regarding, 9
   moral permissibility of, 7, 8, 9, 10, 210, 226n7
   religious views regarding, 10, 210
   See also marriage
Shusterman, Richard, 100–101, 118
Singer, Peter, 126
slavery, 123, 126
sociality, 56–65, 80–82, 85–87, 95, 125–29, 208, 229n4
   complex/political, 62–65, 85–87, 128

   as origin of values, 56–65, 125–29 (see also values: culturally situated)
sociopathy, 230n7
Socrates, 66
soul, 8. See also dualisms: mind/body
Spinoza, Baruch, 79
Sripada, Chandra, 157–59
Stuhr, John, 106
subjectivity. See dualisms: subjective/objective; experience; intersubjectivity; personal identity
Sullivan, Maureen, 9
sympathy, 61. See also empathy

technology, 227n3
   Dewey's conception of, 227n3
   moral knowledge as, 227n3
teleology, 207–8
Thagard, Paul, 66–67
theories of value, naturalistic (defined), 15. See also ethical naturalism
theories of value, non-naturalistic (defined), 15–20. See also Kant, Immanuel: Kantian moral theory; moral fundamentalism; religion, as source of values
theory of mind, 57–58
   as component of moral faculty (Hauser), 147
Tolstoy, Leo, 12
Trevarthen, Colwyn, 57
truth
   found vs. made (Rorty), 236n1
   self-evident (see self-evidence)
truthfulness, 64
Tucker, Don, 55, 149–50
Turiel, Elliot, 35–36

universality
   competing conceptions of, 157–59
   of predispositions, 158–59
   of values, as insufficient evidence of moral faculty/moral grammar, 157–59 (see also values: cross-cultural; values: universal)
utilitarianism. See consequentialism

valuation (reflective), 91, 93, 197, 212, 214, 219
   moral deliberation as a form of, 93–111, 214

INDEX

value, intrinsic vs. extrinsic, 50, 52
values, 1, 49–72, 83–86, 92, 106–7, 114, 121–22, 125–29, 131–32, 135, 138, 141, 143, 149, 156–61
  affective foundations of, 83–85
  American midwestern, 5–6, 8
  biological origins of, 53–57, 77, 78, 84, 131, 138, 149, 159, 192 (*see also* organism-environment interaction)
  and the body, 1, 52–56
  cross-cultural, 6, 35, 49, 59, 130, 131, 156, 157, 160, 161, 186 (*see also* values: universal)
  culturally/historically situated, 1, 35–38, 59, 63, 69, 105–7, 121–22, 125–29, 131, 158, 172, 192
  democratic, 122
  genealogy of, 105–7, 125–29, 132 (*see also* "ethical project")
  innateness, 50–51, 71; of moral principles, 158 (*see also* faculty psychology: moral faculty; instinct; moral grammar)
  "maturationally natural," 51–52, 143 (*see also* organism-environment interaction)
  origins of (*see* organism-environment interaction; meaning: human need for)
  prioritization of, 37–38, 118, 120, 121, 132, 135
  religious (*see* religion: as source of values)
  universal, 55–56, 64–65, 126, 130, 131, 141, 156–59, 161; cultural variations among, 158; as insufficient evidence for moral grammar/moral faculty, 157, 159, 160; as insufficient support for moral realism, 186; possible evolutionary basis of, 138–39 (*see also* faculty psychology: moral faculty; instinct: moral instinct; moral grammar; universality)
  See also sociality: as origin of values; theories of value
valuing (direct/intuitive), 52, 91, 92, 93, 197, 219
virtues, 31–33, 61, 63–69, 219, 229n3, 230n9
  social virtues, 64–65
  virtues conducive to the pursuit of meaning, 67–69
  See also authority: acknowledgment of, as social virtue; character; civic-mindedness; conscientiousness; courage; integrity; loyalty; truthfulness; values

Waal, Franz de, 60–61
war, 12, 104, 226n9
  causes of (in human nature), 154
  drone warfare, 104
  moral permissibility of, 12, 104
Warnock, G. J., 20, 183
will, 23, 24–25, 45, 158, 215
  as collection of habits (Dewey), 25
  heteronomy of (Kant), 45, 158 (*see also* autonomy)
  objectivist conception of, 215
  See also free will
Williams, Bernard, 118
Williams, Robert, 133–35, 152–53, 161
Winter, Steven, 120
Wright, Gary, 228n5

Young, Liana, 152, 158

261